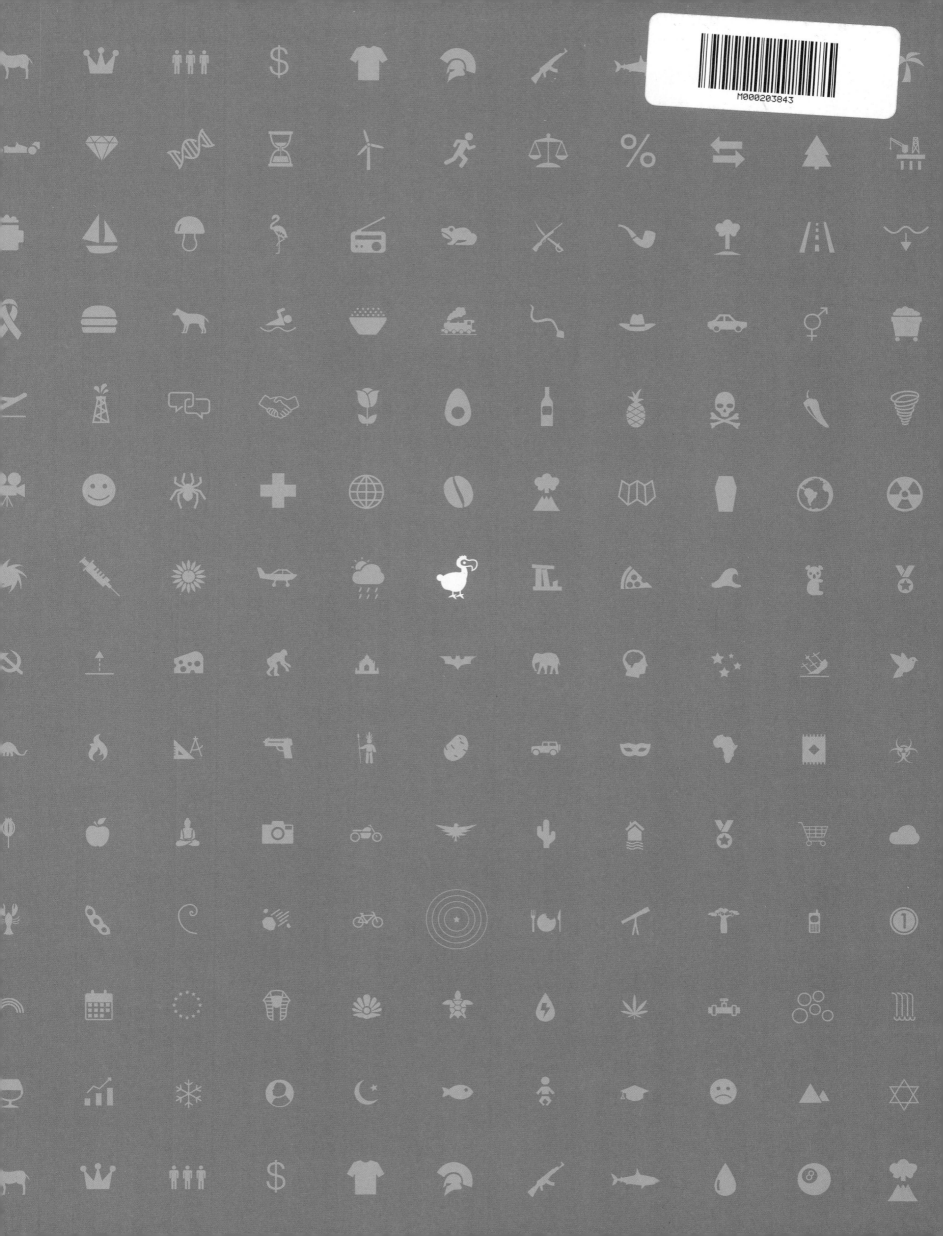

FOR THE INCURABLY CURIOUS

EXPLORER'S ATLAS

Piotr Wilkowiecki & Michał Gaszyński

Published by Collins
An imprint of HarperCollins Publishers
Westerhill Road
Bishopbriggs
Glasgow G64 2QT
www.harpercollins.co.uk

First published 2017

A catalogue record for this book is available from the British Library

ISBN 978-0-00-825305-9

10 9 8 7 6 5 4 3 2 1

Printed in Germany

MIX
Paper from
responsible sources
FSC
www.fsc.org
FSC™ C007454

This book is produced from independently certified
FSC™ paper to ensure responsible forest management.

For more information visit: www.harpercollins.co.uk/green

Front cover image © lynea/Shutterstock.com

If you would like to comment on any aspect of this book, please contact
us at the above address or online.
e-mail: collins.reference@harpercollins.co.uk

 facebook.com/collinsref

@collins_ref

All mapping in this atlas is generated from Collins Bartholomew
digital databases.
Collins Bartholomew, the UK's leading independent geographical
information supplier, can provide a digital, custom, and premium
mapping service to a variety of markets.
For further information:
Tel: +44 (0)208 307 4515
e-mail: collinsbartholomew@harpercollins.co.uk
or visit our website at: www.collinsbartholomew.com

LEGEND

- **●**   CAPITAL CITY

- **○**   OTHER CITY

- **★**   POINT OF INTEREST

- ·············   LINE/ROUTE OF INTEREST

- ————   INTERNATIONAL BOUNDARY

- ·—··—·   DISPUTED BOUNDARY

- ··········   CEASEFIRE LINE

-    RIVER

-    ICE

-    AREA OF INTEREST

-    FOREST

-    MOUNTAINS

- **▲**   SUMMIT

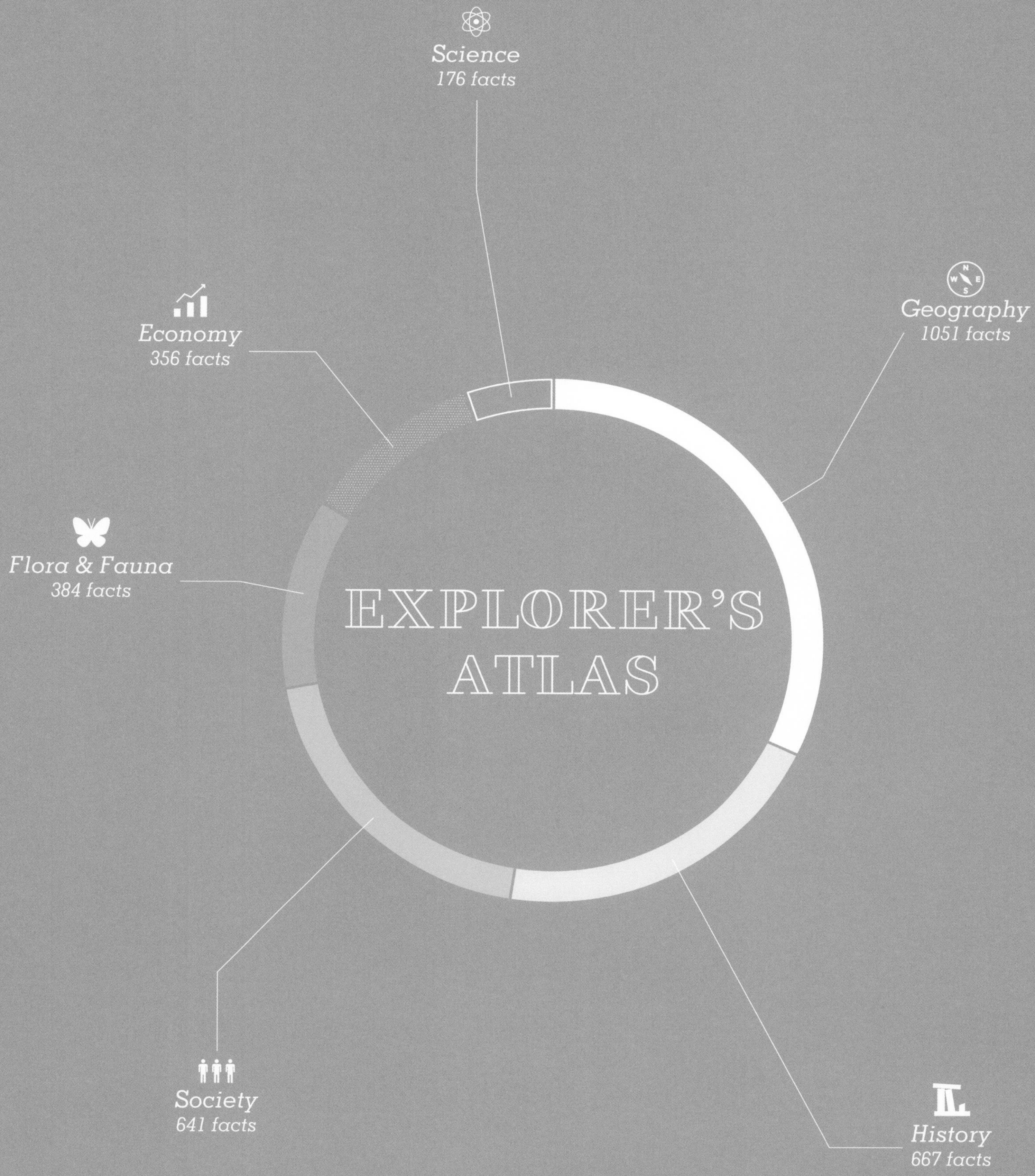

EXPLORER'S ATLAS

Science
176 facts

Geography
1051 facts

Economy
356 facts

Flora & Fauna
384 facts

Society
641 facts

History
667 facts

PLANET EARTH

Earth in the Solar System
Earth is the largest and densest of the four terrestrial planets (Mercury, Venus, Earth and Mars) in the Solar System, made up mostly of rock and metal

Iron planet

Iron constitutes nearly 1/3 of Earth's total mass and 88.8% of its core. Currents flowing in the outer core generate Earth's geomagnetic field which protects all life from powerful solar winds and cosmic radiation

Earth's surface

Land 29%

Water 71%

96.5%
Salt water

3.5%
Fresh water

Artificial satellites

Currently there are around 3,600 artificial satellites in orbit of which over 1,000 remain operational

Just a fraction
Although seas and oceans constitute most of our planet's surface, water makes up only 0.02% of its mass

_ _ _ _ Tropic of Cancer

Tropic of Cancer
Northernmost location where the Sun can appear directly overhead. It passes through 16 countries

Earth core's life
Some ancient civilizations believed there was another world inside our planet

Extinction
99% of all species that ever lived on Earth are now extinct

There have been 5 major extinction events. The last one, 66 million years ago, caused by the impact of a comet or asteroid or volcanic activity, ended the reign of the dinosaurs

Around 90% of South America lies in the Southern Hemisphere

Distance to Moon
384,400 km or 1.28 light-seconds

Moon

Solar System's largest satellite (in relation to its primary planet). It is responsible for ocean tides and stabilizing the Earth's rotational axis

Low Earth orbit

A low Earth orbit has an altitude between 0 and 2,000 km. Currently there are around 500 operational satellites in this orbit. Their orbital period is between 84 and 127 minutes

Arctic Circle

The area north of the Arctic Circle covers around 4% of the surface of the globe and it is populated by around 4 million people

Medium Earth orbit

Altitude between 2,000 km and 35,786 km. It is used by the Global Positioning System (GPS)

Brandt Line

30° N is generally considered the north-south division between the economies of the world with the North, plus Australia and New Zealand, controlling around 80% of global income

Geostationary orbit

Directly above the equator at an altitude close to 35,786 km. Satellites here have an equal orbital period to the Earth's rotational period

Approximately 2/3 of Africa lies in the Northern Hemisphere

Northern Hemisphere

Contains 67.3% of Earth's land

Most spoken language: Mandarin

Population: 90% of the world's population

Most populous city: Tokyo (38.2 million)

Better for deep-space observation because the Milky Way is not as blinding to the observer

Purple planet

When the earliest life evolved, instead of green chlorophyll there was a different, purple molecule, capable of supporting photosynthesis. This might have resulted in Earth once being purple, just as the land is green today when seen from space

Equator

Imaginary line extending over 40,075 km. It would take the fastest marathon runner 81 days non-stop to run this distance

21.3% of the equator passes over land (of 11 independent countries) and 78.7% over seas and oceans

Southern Hemisphere

80.9% is covered with water

Most spoken language: Portuguese

Most populous city: São Paulo (21.5 million)

Relatively milder climate because it has more water that heats up and cools down slower than land

Cyclones in the Southern Hemisphere spin clockwise, the opposite to those in the North, due to the Coriolis Effect

Prime Meridian

The line of 0° longitude that runs through the Royal Observatory, Greenwich in London, established in 1851. It passes through 8 countries and the Norwegian claim in Antarctica

Earth's diameter

12,742 km, similar distance to Los Angeles to Jerusalem

Earth's diameter

Prime Meridian

Tropic of Capricorn

Southernmost location where the Sun can appear directly overhead. It passes through 10 countries

Gravity and bacteria

In zero gravity, some bacteria, including Salmonella, responsible for food poisoning, mutate and can become three times more malignant than on Earth

Antarctic Circle

At least once per year here, the Sun is above or below the horizon for 24 continuous hours

Earth's speeds

Earth rotates at different speeds in different places, depending on the latitudinal position. It's the fastest at the equator (1,669 km/h). A person standing on the Arctic Circle would rotate just 736 km/h to Earth's axis, and at the North and South Poles, would stand completely still

GREATEST
EXPLORERS

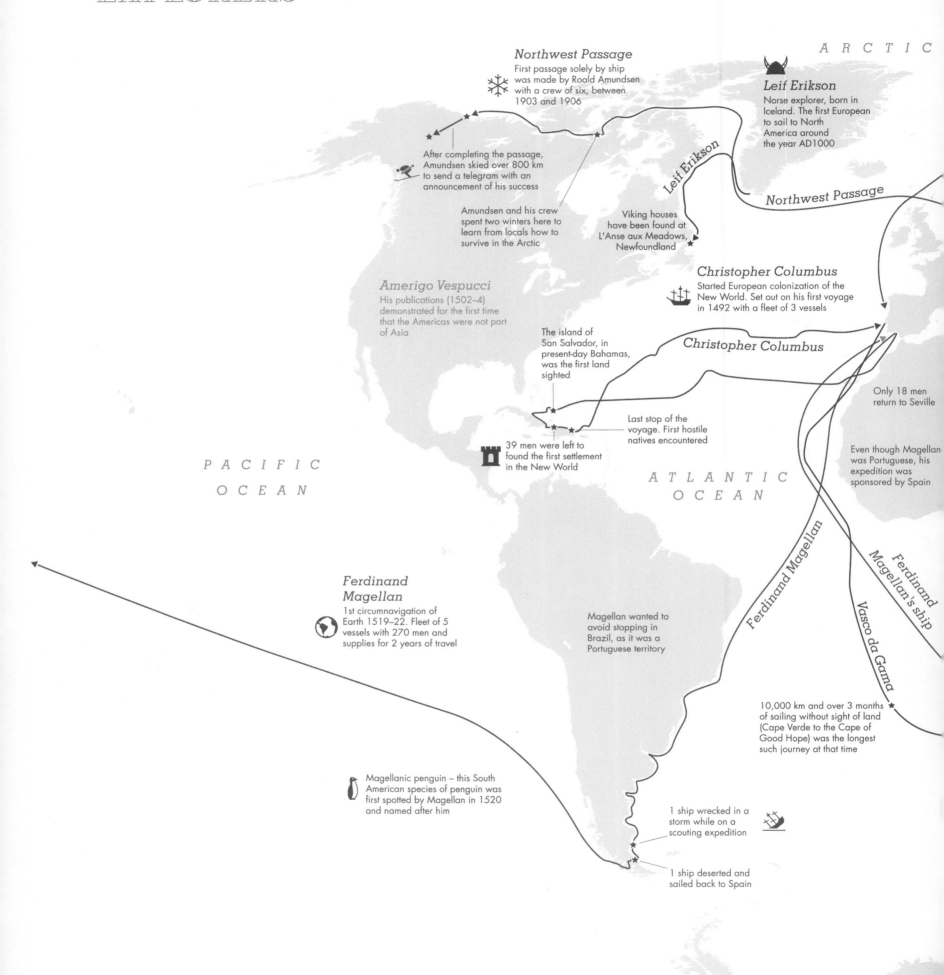

Northwest Passage

First passage solely by ship was made by Roald Amundsen with a crew of six, between 1903 and 1906

After completing the passage, Amundsen skied over 800 km to send a telegram with an announcement of his success

Amundsen and his crew spent two winters here to learn from locals how to survive in the Arctic

A R C T I C

Leif Erikson

Norse explorer, born in Iceland. The first European to sail to North America around the year AD1000

Northwest Passage

Viking houses have been found at L'Anse aux Meadows, Newfoundland

Leif Erikson

Christopher Columbus

Started European colonization of the New World. Set out on his first voyage in 1492 with a fleet of 3 vessels

Amerigo Vespucci

His publications (1502–4) demonstrated for the first time that the Americas were not part of Asia

The island of San Salvador, in present-day Bahamas, was the first land sighted

Christopher Columbus

Last stop of the voyage. First hostile natives encountered

39 men were left to found the first settlement in the New World

*PACIFIC
OCEAN*

*ATLANTIC
OCEAN*

Only 18 men return to Seville

Even though Magellan was Portuguese, his expedition was sponsored by Spain

*Ferdinand
Magellan*

1st circumnavigation of Earth 1519–22. Fleet of 5 vessels with 270 men and supplies for 2 years of travel

Magellan wanted to avoid stopping in Brazil, as it was a Portuguese territory

Ferdinand Magellan

*Ferdinand
Magellan's ship*

Vasco da Gama

10,000 km and over 3 months of sailing without sight of land (Cape Verde to the Cape of Good Hope) was the longest such journey at that time

Magellanic penguin – this South American species of penguin was first spotted by Magellan in 1520 and named after him

1 ship wrecked in a storm while on a scouting expedition

1 ship deserted and sailed back to Spain

0	1,000	2,000	4,000 km

0	500	1,000	2,000 miles

OCEAN

Northeast Passage

Northeast Passage

Portuguese explorer David Melgueiro is considered the first person (1660–2) to cross it in a ship

The first crossing of the Northeast Passage was probably possible thanks to an exceptionally warm period in the 1660s

"The Travels of Marco Polo" book was an inspiration for Columbus's voyages

Marco Polo

A merchant traveller, first person to describe a journey (1271–95) to the Far East in detail. The journey took 24 years and covered almost 24,000 km

Marco Polo

Marco Polo was just 17 years old when he set off with his father and uncle

PACIFIC OCEAN

Magellan died in a battle on the island of Mactan, the Philippines, on 27 April 1521. The remaining two ships sailed on to the Moluccas

Vasco da Gama

Discovery of the sea route to India (1497–9). 4 ships with 170 men

Da Gama failed to secure a commercial treaty with Calicut. His gifts failed to impress the King of Calicut as they lacked gold and silver

1 ship abandoned due to a lack of crew

Expedition looted unarmed ships of Arab merchants

INDIAN OCEAN

1 ship began to take on water and was left behind

On the outgoing journey, the fleet crossed the Indian Ocean in 23 days and on their return, in 132 days

Matthew Flinders

Matthew Flinders

English navigator, the first man to circumnavigate Australia (1802–3) and identify it as a continent

Vasco da Gama pretended to be a Muslim to gain an audience with the Sultan of Mozambique

Before the expedition reached the Cape of Good Hope, only rice was left as rations. 20 men died of starvation

He sighted a French ship here on a similar expedition. Although both countries were at war, they exchanged their discoveries and named the place Encounter Bay

THERN OCEAN

When Flinders was heading back to England, he was arrested in Mauritius and imprisoned for six years where he recorded descriptions of his voyages. He died shortly after reaching home in 1810

EARTHQUAKES & VOLCANOES

Beerenberg (2,277 m)
Northernmost active volcano

Kola Superdeep Borehole
The deepest hole drilled in the Earth's crust is 12,262 m deep

EURASIAN PLATE

1.0

2.3

Eyjafjallajökull
Its 2010 eruption forced almost all European countries to close their air space. The airline industry lost up to $200 million a day

1976 Tangshan earthquake
With 242,000 and possibly up to 700,000 deaths, it is the deadliest earthquake of modern times

1556 Shaanxi earthquake
Considered the deadliest earthquake on record, killing around 830,000 people. Most of the population affected lived in caves which collapsed

"Lighthouse of the Mediterranean"
Stromboli volcano has been erupting almost continuously for the past 2,000 years

2.5

Mid-Atlantic Ridge
This underwater mountain range spreads as fast as your fingernails grow (10–40 mm/year)

ARABIAN PLATE

AFRICAN PLATE

Geothermal power in Kenya
Geothermal energy production in Kenya accounts for about 50% of its total electric capacity

Toba catastrophe
After this eruption, 75,000 years ago, a 15 cm layer of ash covered the entire region of Southeast Asia. It is the largest super-eruption on Earth of the past 2.5 million years

3.0

"The year without a summer"
Ash from the 1815 eruption of Mount Tambora caused a "dry fog" which reduced the power of the sun, leading to the collapse of crop production around the planet. In Europe and Asia it led to famines and epidemics

4.4

3.5

1.4

Stratovolcano
Type of volcano with a steep, conical shape and producing explosive eruptions. They form when tectonic plates collide, releasing energy from the Earth's mantle

3.5

SOUTH AMERICAN PLATE

7.2

1.4

7.5

SCOTIA PLATE

Submarine volcanoes
Most of the magma output on Earth is from submarine volcanoes. Their number is estimated at over 1 million

ANTARCTIC PLATE

Plate tectonics
The uppermost region of the Earth's mantle, 100 km thick, floating on an ocean of viscous, molten rock, is called the asthenosphere. It is involved in plate tectonic movement

| 0 | 1,000 | 2,000 | | 4,000 km |
| 0 | 500 | 1,000 | | 2,000 miles |

Global warming myth
Contrary to popular myth, people emit 100 times more CO_2 than volcanoes

Geothermal power in the US
Although the US is the world's leader in geothermal electric capacity with 3,450 MW, it accounts for only 0.3% of its total electricity generation

1958 Lituya Bay megatsunami
The largest tsunami ever recorded, caused by a rockslide triggered by an earthquake. Waves up to 30 m high reached 525 m above the height of the bay

1906 San Francisco earthquake
One of the deadliest natural disasters in the history of the US killed over 3,000 people and destroyed 80% of the city

San Andreas Fault
As the two tectonic plates move against each other, San Francisco and Los Angeles are getting closer to each other by 20 to 35 mm every year

NORTH AMERICAN PLATE

Most earthquakes
Japan is the country with the highest density of earthquakes, up to 1,500 per year. The first reliably documented earthquake took place here in AD 599

Hawai'i island
The largest Hawaiian island is formed from five volcanoes that overlap each other

PHILIPPINE PLATE

CARIBBEAN PLATE

COCOS PLATE

8.6

5.0

Smallest volcano
World's smallest volcano, 3 cm high, has been discovered in Colombia

SOUTH AMERICAN PLATE

Volcanic capital
Of all the countries in the world, Indonesia has the greatest number of historically active volcanoes

PACIFIC PLATE

Fastest plates
The fastest moving plates are the Nazca and Cocos, up to 150 mm/year, about as fast as your hair grows

Ring of Fire
The area which encircles the Pacific Ocean basin where 75% of the world's active and dormant volcanoes are located and where 90% of earthquakes on the planet occur

15.1

NAZCA PLATE

Nevado Ojos del Salado
World's highest active volcano (6,908 m)

15.1

Tsunami
A tidal wave usually resulting from seismic activity. They can travel across oceans at speeds of up to 805 km/h, reaching extreme heights as they reach the coast. Minutes before a tsunami strikes a drawback can occur at the shoreline, the water retreating up to hundreds of metres

INDO-AUSTRALIAN PLATE

Great Chilean earthquake
The 1960 earthquake is the most powerful ever recorded with a magnitude of 9.5

DART
Warning system consisting of 41 buoys which detect changes in pressure in the deep ocean caused by tsunamis

9.4

5.9

SCOTIA PLATE

Mount Erebus
World's southernmost active volcano and the only one where the lava lake is visible over long periods of time

- - - - - - Destructive boundary

――――― Conservative boundary

━━━━━ Constructive boundary

2.4 ――→ Direction and rate of movement (cm per year)

HIGHEST MOUNTAINS

There are at least 109 mountains higher than 7,200 m, most of them located in the Himalaya and Karakoram ranges

The highest mountain outside Asia, Cerro Aconcagua (6,961 m), is only the 189th highest on Earth

Eight-thousanders

There are 14 mountain peaks that are more than 8,000 m above sea level. All of them are located in the Himalaya and Karakoram ranges

	Height	Location
Mount Everest	8,848 m	China/Nepal
K2	8,611 m	China/Pakistan
Kangchenjunga	8,586 m	India/Nepal
Lhotse	8,516 m	China/Nepal
Makalu	8,463 m	China/Nepal
Cho Oyu	8,201 m	China/Nepal
Dhaulagiri I	8,167 m	Nepal
Manaslu	8,163 m	Nepal
Nanga Parbat	8,126 m	Pakistan
Annapurna I	8,091 m	Nepal
Gasherbrum I	8,068 m	China/Pakistan
Broad Peak	8,047 m	China/Pakistan
Gasherbrum II	8,035 m	China/Pakistan
Xixabangma Feng	8,027 m	China

First 14 ascents

In 1986, Italian Reinhold Messner became the first person to climb all eight-thousanders. He achieved this without the aid of oxygen

First eight-thousander ascent

First recorded successful ascent of an eight-thousander was in 1950 when Frenchmen Maurice Herzog and Louis Lachenal climbed Annapurna I

Most eight-thousanders climbed

Phurba Tashi of Nepal has climbed eight-thousanders 30 times, more than anyone else

Most successful country

Italy has the highest number of climbers who have ascended all eight-thousanders with 7

Olympus Mons and Rheasilvia

The highest known mountains in the Solar System are Olympus Mons on Mars (23 km) and the central peak of Rheasilvia, a crater located on the asteroid Vesta, which is of a comparable height

Mauna Kea
Hawaii, USA
4,205 m

Earth's highest mountain measured from its ocean floor base

Sea level

6,000 m

Seven Summits

Each continent's highest mountain. Summiting all of them is a famous mountaineering challenge. It was first achieved by an American businessman Richard Bass in 1985. However, he chose Mount Kosciuszko as the highest peak in Oceania

One year later, Canadian Patrick Morrow achieved this feat, climbing Puncak Jaya instead, which is defined as Oceania's highest mountain, according to the Messner list

Mount Everest
Asia, China/Nepal
8,848 m

Earth's highest mountain but not the most difficult to climb. 641 people made the summit in 2016

Cerro Aconcagua
South America, Chile
6,961 m

The highest mountain in the Western and Southern Hemispheres

Denali (Mount McKinley)
North America, USA
6,190 m

3rd most topographically prominent and isolated summit

Kilimanjaro
Africa, Tanzania
5,892 m

Highest volcano outside South America

El'brus
Europe, Russia
5,642 m

West summit first ascended in 1874, first of the Seven Summits

Mount Vinson
Antarctica
4,897 m

First spotted in 1958 by a US Navy aircraft

Puncak Jaya
Oceania, Indonesia
4,884 m

World's highest island peak

LONGEST RIVERS

Amacayacu National Park
Home to the Ticuna people, the most populous indigenous tribe in the Brazilian Amazon

2. Amazon
South America's longest river (6,516 km). Some sources classify it as the world's longest

The discharge of the Amazon constitutes 20% of all global river flow. It could fill the Dead Sea in less than a week

ATLANTIC OCEAN

Marajó
About the size of Switzerland, formed from the sediments carried by the river

MANAUS
Most populous city (2 million) on the Amazon. Due to its remoteness, it's accessed primarily by boats or planes

Allpahuayo Mishana National Reserve
One of the most biodiverse spots on Earth. 500 different varieties of tree can be found per hectare

7. Huang He (Yellow River)
World's 7th longest river (5,464 m)

The river is called "yellow" because of its muddy water

BO HAI

The three deadliest floods in history (1887, 1931, 1938) were on the Yellow river and killed up to 6.8 million people

GULF OF OB'

5. Ob'-Irtysh
2nd longest river system in Asia (5,568 km)

Northern river reversal
Soviet plan to divert the flow of the Siberian rivers towards Central Asia's agricultural areas that lack water

OMSK
Western Siberia's largest river port

MEDITERRANEAN SEA

Rosetta Stone
The key to deciphering Egyptian hieroglyphs was found here in 1799

CAIRO
Population: 19.5 million

Before the last glacial period, The Nile flowed into the Gulf of Sirte, 1,000 km to the west

Stone blocks used for the construction of the Pyramid of Khufu were transported from here, 800 km by water

1. Nile
World's longest river (6,695 km)

Egyptian civilization's development resulted greatly from its ability to use the Nile and its annual floods for agriculture

Blue Nile
White Nile
Nile

The Nile's drainage basin covers 11 countries and around 10% of Africa

Location of the source of the Nile is still disputed

Until the 19th century, the Missouri was believed to be part of the Northwest Passage connecting the Pacific and Atlantic oceans

A drop of water from the Mississippi's headwaters has to travel 90 days to reach the Gulf of Mexico

Many of the river's meanders have been cut to ease navigation which has reduced its length by 320 km

MINNEAPOLIS
First bridge across the Mississippi was built here in 1855

Missouri
Mississippi
ST LOUIS

4. Mississippi-Missouri
Longest river system in North America (5,969 km)

The Mississippi formed much of the boundary of the United States as set out in the 1783 Treaty of Paris

EAST CHINA SEA

3. Chang Jiang (Yangtze)
Longest river in Asia (6,380 km)

Geladaindong Peak (6,621 m)
Its glaciers are the source of the river

Its basin accounts for 75% of China's total rice production

Three Gorges
Series of beautiful gorges with spectacular scenery

SHANGHAI
Population: 25.2 million

GULF OF YENISEY

6. Yenisey-Angara-Selenga
Yenisey-Angara-Selenga is the 3rd longest river system in Asia (5,550 km)

During WWII, Hitler and the Japanese had plans to divide Asia along the river

The Taimyr reindeer herd, the world's largest (around 600,000), migrates along the river

Its entire length was first navigated by an Australian-Canadian team in 2001

GULF OF MEXICO

8. Congo
One of the world's deepest rivers (220 m), 2/3 of the height of the Eiffel Tower

BRAZZAVILLE
Capital of Congo

2nd longest river in Africa - 4,667 km

ATLANTIC OCEAN

KINSHASA
Capital of the Dem. Rep. of the Congo

Approximate location of the deepest point of the river

Congo Canyon
Created by the river, it is one of the world's largest submarine canyons

2nd largest river by discharge. If it was directed into Lake Chad, it would replace its current water level in just 20 days

21. Murray-Darling
Oceania's longest river system (3,672 km)

The Darling river suffers from water overuse, pollution and droughts, and in some seasons barely flows

The Murray-Darling basin drains about 1/7 of Australia's land and is the most significant agricultural area in the country

The majority of the Volga freezes for about 3 months of the year

Moscow Canal
128 km long. Connects the capital with the Volga

MOSCOW

20. Volga
Europe's longest river (3,688 km)

World's longest river, which flows to an enclosed basin

VOLGOGRAD
Formerly Stalingrad

60% of the Volga's annual drainage comes from melting snow

The Volga's delta is one of the main centres of the caviar industry

CASPIAN SEA

500   1,000   2,000 km

300   600   1,200 miles

LARGEST COUNTRIES

The European part of Russia accounts for around 23% of the country's total area and 77% of its population

1. Russia
Area: 17,075,400 km^2

Population: 143,457,000

The largest country. It has an area only slightly smaller than Pluto (17,790,000 km^2)

The Canadian Arctic Archipelago itself makes up around 14% of Canada's total area

2. Canada
Area: 9,984,670 km^2

Population: 35,940,000

Has the largest water area of all countries – 891,163 km^2

Alaska is the largest US state and accounts for 17% of the country's area

3. United States of America
Area: 9,826,635 km^2

Population: 321,774,000

If not counting the state of Alaska, the US would be smaller than Brazil

Hawaii is the 8th smallest US state and accounts for 0.2% of the country's area

The US would be the 2nd largest country in the world if only land surface was counted and areas of water were excluded

China is the world's largest exporter of goods, worth over $2 trillion per year

At 22,147 km, China has the longest land border of all countries

4. China
Area: 9,606,802 km^2

Population: 1,383,925,000

If considering only land area, China is the 2nd largest country after Russia

China is the greatest emitter of carbon dioxide, 10.3 billion metric tons per year

5. Brazil
Area: 8,514,879 km^2

Population: 207,848,000

Largest contiguous territory in the Americas

Brazil covers 47% of South America

The largest Portuguese-speaking country

6. Australia
Area: 7,692,024 km^2

Population: 23,969,000

The largest country with no land borders and the largest that lies completely in the Southern Hemisphere

Australia has the largest territorial claim in Antarctica of all countries – 5.9 million km^2

7. India
Area: 3,166,620 km^2

Population: 1,311,051,000

The most populous democratic country in the world

India has the 2nd largest active army – 1,395,100 military personnel

The Deccan Peninsula is the second largest peninsula in the world

8. Argentina
Area: 2,766,889 km^2

Population: 43,417,000

Although Argentina has an area over 5 times larger than Spain, it has roughly the same population as its colonizer

The largest Spanish-speaking country

9. Kazakhstan
Area: 2,717,300 km^2

Population: 17,625,000

Largest landlocked country

The largest country of Central Asia. It generates around 60% of the region's GDP

10. Algeria
Area: 2,381,741 km^2

Population: 39,667,000

Largest country in Africa

Only 1% of its land area is covered by forest

0	500	1,000		2,000 km

0	250	500		1,000 miles

SMALLEST COUNTRIES

1. Vatican City

Area: 0.5 km²

Population: 800

The only undisputed independent state which is not a member of the United Nations

2. Monaco

Area: 2 km²

Population: 38,000

World's most densely populated country (19,000 people per km²)

3. Nauru

Area: 21 km²

Population: 10,000

The smallest country in the Pacific

4. Tuvalu

Area: 25 km²

Population: 9,916

Only 7 times larger than Central Park in New York

5. San Marino

Area: 61 km²

Population: 32,000

The only country that has more motor vehicles than people – 1.263 per person

6. Liechtenstein

Area: 160 km²

Population: 38,000

Together with Uzbekistan, they are the only double landlocked countries

7. Marshall Islands

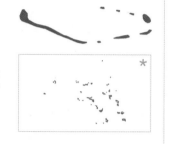

Area: 181 km²

Population: 53,000

Although the country's area is about three times the size of Manhattan Island, it is spread across 1,156 islands

8. St Kitts and Nevis

Area: 261 km²

Population: 56,000

Smallest country in the Western Hemisphere, by both area and population

9. Maldives

Area: 298 km²

Population: 364,000

The smallest Asian country by area and population

If the current trend of sea level rise continues, the country will become submerged in 30 years

10. Malta

Area: 316 km²

Population: 419,000

Smallest and most densely populated country in the European Union

11. Grenada

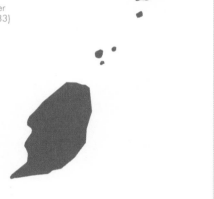

Area: 378 km²

Population: 107,000

The smallest country ever invaded by the US (1983)

In 2004, Grenada's first hurricane in 49 years damaged or destroyed 90% of country's buildings

13. Barbados

Area: 430 km²

Population: 284,000

Roughly the same GDP as Mauritania which is over 2,400 times larger

12. St Vincent and the Grenadines

Area: 389 km²

Population: 109,000

Roughly the same size as Las Vegas

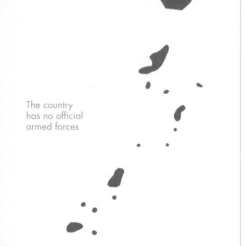

The country has no official armed forces

14. Antigua and Barbuda

Area: 442 km²

Population: 92,000

Called "Land of 365 Beaches" because of the many beaches that surround the islands

15. Seychelles

Area: 455 km²

Population: 96,000

The smallest African country by both population and area

Only gained independence from the UK in 1976

Highest nominal GDP per capita of all African countries

16. Andorra

Area: 465 km²

Population: 70,000

With an average elevation of 1,996 m, it is the 6th highest country in the world

* Whole island group shown at a smaller scale

| 10 | 20 | 40 km |

| 5 | 10 | 20 miles |

Sunflowers
North Dakota grows around half of the US's sunflowers. It is estimated that blackbirds eat up to $10 million worth of them every year

NORTH AMERICA

ARCTIC OCEAN

Cape Prince of Wales
Westernmost point in mainland North America
168°5'5" W

Zenith Point
Northernmost point in mainland North America
72°0'0" N

Human migration
There is a theory that the first humans reached North America between 40,000 and 17,000 years ago, across the Bering land bridge

Yukon

Great Divide
The line connecting the highest peaks of the Rocky Mountains divides the continent into the Pacific and Atlantic drainage regions

NORTH AMERICA

Area
North America has an area of 24,680,331 km². It's the 3rd largest continent, occupying 16.6% of the Earth's land surface and 4.8% of its total surface area

Population
North America's most populous countries are: United States (321 million), Mexico (127 million) and Canada (36 million)

North America (7.79%)
579,024,000 people

Exxon Valdez oil spill
One of the most devastating environmental disasters caused by humans. In 1989, around 260,000 barrels of crude oil spilled from the oil tanker, covering around 2,100 km of coastline. The immediate effects included the deaths of up to 250,000 seabirds and an unknown number of other animals

23 countries
There are 23 countries included in the North American continent and many non-sovereign territories of UK, USA, France, Netherlands, Denmark and Colombia

Port of Vancouver
The largest port on the West Coast of America

Wealth
The nominal GDP of the continent is $21.2 trillion, 2nd only to Asia. Around 87% of it is generated by the US

Mastodon
Species distantly related to the elephant that became extinct 10,000 to 11,000 years ago. Up to 3.25 m tall and 11 tonnes in weight, it inhabited North and Central America. The largest elephants alive today weighs around 7 tonnes

Great Divide

Rocky Mountains

Death Valley
Continent's lowest point at -86 m

● LOS ANGELES
Population: 12.3 million

HAWAI'IAN ISLANDS

Isolation
Around 3,600 km from contiguous USA

Middle America
Region comprising Mexico, Central America, the Caribbean and sometimes Colombia and Venezuela

PACIFIC OCEAN

Maya civilization
Developed a writing system, a calendar and a concept of zero around 500 years before the Old World civilizations

0	500	1,000		2,000 km
0	300	600		1,200 miles

GREENLAND
The world's largest island is geographically part of the continent because it lies on the same tectonic plate – North Amercian Plate

Aurora
These high latitude green lights are electrons and protons from the solar wind striking the Earth's upper atmosphere. They occur mainly in a band between 3° to 6° wide in latitude and 10° to 20° from the geomagnetic poles

Polar bears
Most of the world's polar bears live in Canada. Their fur is transparent, their skin completely black and below it they have a 12 cm layer of fat

Nettilling Lake
With an area of 5,066 km², it is the world's largest lake on an island

Jacques Cartier
French explorer (1491–1557) who claimed the territory of what is now Canada for France

Cape St Charles
Easternmost point in mainland North America 55°37'15"W

Samuel de Champlain
French explorer (c.1567–1635) who mapped much of northeastern North America. In 1608, he established a settlement which later became the city of Quebec

Oldest rocks
The Hudson Bay area of Canada is the world's oldest region as some of its rocks are over 4 billion years old

Geographic centre
Rugby, North Dakota is the geographic centre of the continent

Trade
Trade relations between the US and China are the largest in the world followed by those between the US and Canada

Lion's mane jellyfish
The largest jellyfish ever (37 m long) was found in Massachusetts Bay in 1870

Richest pirate
The first authentic pirate ship ever discovered in North America, sank here in 1717. It belonged to Captain Samuel "Black Sam" Bellamy, the richest pirate in history

NEW YORK
Population: 18.6 million

Mississippi

Missouri

Appalachian Mountains

Great Divide Basin
Wyoming's Red Desert is an endorheic basin which means that none of the rainfall drains into an ocean

Universalis Cosmographia
Published in 1507 by Martin Waldseemüller, it was the first world map that showed the Americas separate from Asia

ATLANTIC OCEAN

BERMUDA
Despite lying so close to the US, from a geological point of view, this British territory is not part of the North American continent

Dinosaurs
The largest variety of dinosaurs have been found in the US

Amerigo Vespucci
The Americas were named after Amerigo Vespucci – an Italian explorer who was the first person to state that this land mass was not the East Indies

Isthmus of Tehuantepec
Recognized by some geologists as a border between the Americas. Before the opening of the Panama Canal, it was a major trade route between the oceans

Alternative route
There was a time when Central America was underwater and the Americas were connected via a land bridge along the West Indies

North or South America?
Aruba, Curaçao, Bonaire and Trinidad and Tobago lie on the South American continental shelf but are generally considered part of North America

MEXICO CITY
Population: 21.3 million

Great Divide

Aztec Empire
Before the 1521 Spanish conquest, it was the most dominant civilization of Middle America

Punta Mariato
Southernmost point in mainland North America 7°12'32" N

CHUKCHI SEA

ARCTIC OCEAN

QUEEN ELIZABETH ISLANDS
This archipelago contains around 14% of all continental glaciers (excluding Greenland and Antarctica)

Population
Alaska has the lowest population density of all the US states (0.49/km²) and the 48th largest population (738,432 people)

Alaska Purchase
The US bought Alaska from Russia in 1867 for $7.2 million

BEAUFORT SEA

Coastline
Alaska's coastline is longer than all other US states' coastlines combined

Yukon
3rd longest river in the US (3,185 km)

Oil & gas
Alaska's most important revenue source is its oil and natural gas. The state accounts for 25% of the oil produced in the US

Forests
Around 42% of the country is forested which accounts for 10% of the world's forests

Great Bear Lake

BERING SEA

Size
The largest US state, similar in size to Libya. The state of Rhode Island would fit here 429 times

USA Alaska

Denali (6,190 m)
The highest mountain in North America. Before 2015 it was named Mt McKinley

Canadian diamonds
Mined in northwestern Canada, they were promoted as an alternative to those imported from Africa to eliminate the risk of purchasing blood diamonds

Low gravity
A large area of northern Canada has unusually low gravity. This is because glacier activity during the Ice Age carved the crust and then with the ice melting so rapidly, the crust hasn't yet fully "bounced back"

Wrangell-St Elias National Park
The largest US national park (53,321 km²), similar in size to Croatia

WHITEHORSE
The only city in the territory of Yukon

PACIFIC OCEAN

Tongass National Forest
Temperate rainforest and the largest national forest in the US (69,000 km²), similar in size to Ireland

Ice hockey
The official national winter sport of Canada, but in 1920, it was also included in the Summer Olympics in Antwerp

Great Slave Lake
Deepest lake in North America (614 m)

CAN

Slave

Lake Athabasca

Rocky Mountains

Mackenzie

Athabasca oil sands
Alberta has a proven reserve of 170 billion barrels, the world's 2nd largest reserves of crude oil. Extracting it has a negative impact on the environment (water waste, CO₂ emissions, soil erosion and river pollution)

Immigration
20% of Canadians were born outside Canada

Lake Winnipegosis

Coastline
British Columbia has over 25,725 km of coastline

Athabasca

Lakes
Canada has 31,752 lakes larger than 3 km². The overall number of lakes is almost impossible to determine

Canadian provinces and territories

Yukon
Northwest Territories
Nunavut
Newfoundland and Labrador
British Columbia
Alberta
Saskatchewan
Manitoba
Ontario
Québec
Prince Edward Island
Nova Scotia
New Brunswick

WHISTLER BLACKCOMB
The largest ski resort in North America

CALGARY
"The cleanest city in the world"

75% of Canada's population lives within 160 km of the Canada/US border

VANCOUVER
Often described as one of the most beautiful cities in the world with a very high quality of life

Longest border
The border between Canada and the US is the world's longest border between two countries at 8,891 km

0   200   400   800 km

0   125   250   500 miles

Nares Strait
At its narrowest point, Greenland and Canada are roughly 20 km apart

During 1962–4, an iceberg three times larger than Manhattan drifted through the strait to the Atlantic

ELLESMERE ISLAND
World's 10th largest island

BAFFIN BAY

East Greenland Current
The current is important as it connects the Arctic to the North Atlantic, carrying sea ice and creating the largest freshwater sink for the Arctic

BAFFIN ISLAND
5th largest island, similar in size to Spain with a population of around 11,000 people

East Greenland Current

LABRADOR SEA
Labrador Sea provides a significant part of the North Atlantic Deep Water which is the largest identifiable water mass in the world's oceans

ATLANTIC OCEAN

Newfoundland & Labrador province
94% of the province's population live in Newfoundland

L'Anse aux Meadows
First evidence of Norse settlement in North America

Education
Over 50% of Canada's population hold college degrees

"Li"
The most common surname in Canada

Voting for independence
In 1995, Québec almost voted for independence and separation from Canada. The referendum failed by less than 1%

Time zone
Newfoundland & Labrador province has its own time zone, UTC-3:30

NEWFOUNDLAND

HUDSON BAY
2nd largest bay in the world

Henry Hudson
English explorer who sailed to northern North America 4 times in order to find a route to Asia. Died with his son in 1611 after his crew mutinied. He began the Dutch colonization of this region

Québec
In 2016, nearly 71% of the world's maple syrup was produced in this province

Reindeer Lake

Manicouagan Reservoir
Probably created over 200 million years ago when a 5-km-wide meteor hit the Earth

Prince Edward Island
The smallest Canadian province, 10 times smaller than the 2nd smallest, Nova Scotia

Larger than China?
World's 2nd largest country by total area, but if one excludes 1.5 million km² of the country's internal waters, China is bigger

Québec hotels
Château Frontenac is the most photographed hotel in the world

Hôtel de Glace is the first ice hotel in North America. Every year, it takes 45 days to rebuild it

St Lawrence river estuary is the largest in the world

Population
Canada's population is similar in size to that of the Greater Tokyo Area, however Canada's territory is around 770 times larger

French language
Mother tongue of over 7 million out of 35 million Canadian citizens

World's 4th largest French-speaking city

QUÉBEC

St Lawrence

Yonge Street
The longest street in the world (1,896 km)

MONTRÉAL

OTTAWA

"Dutch room"
During WWII, Dutch Princess Margriet was born in Ottawa Civic Hospital. The Governor General of Canada had to declare the Princess's room "extraterritorial" so that she'd have Dutch nationality exclusively and still be in line to the throne

Lake Winnipeg

Lake Manitoba

Lake Superior

L. Ontario

TORONTO

Niagara Falls
51 m high. It is the largest waterfall (that has a vertical drop higher than 50 m) by flow rate in the world – 2,400 m³/s. In 2007 it was visited by 20 million people

Lake Huron

Lake Michigan

Lake Erie

Trade partners
Canada is the main importer of US goods

Fort Clatsop

Lewis & Clark Expedition

NEWPORT

US Route 20
The longest road in the country. The "0" indicates that it is a coast-to-coast route

US Route 20

Trade partners
China is the main exporter to the US. The main importer of US goods is Canada

Railways and roads
USA has the largest network of railways (224,792 km) and roads (6,586,610 km)

Pacific Railway

SACRAMENTO

Pacific Railway
First transcontinental railroad, finished in 1869. Construction started at both ends and met here, in Promontory, Utah

Navajo Nation
The largest Native American reservation in the US, comparable in size to West Virginia

Mount Whitney (4,418 m)
The highest peak in the contiguous US

Internet
First message sent through the internet was from University of California, Los Angeles (UCLA) to Stanford Research Institute (SRI), in 1969

Route 66

LOS ANGELES

Miss Universe
USA has won the contest 8 times, more than any other nation. Additionally, Puerto Rico has won 5 times

Largest airline
American Airlines Group is the world's largest private airline company by revenue ($41 billion) and profit ($7.6 billion)

Missouri

Lewis & Clark Expedition
First attempt to explore the West (1804–6). Sacagawea, a Shoshone woman who served as an interpreter for Lewis and Clark, is featured on the $1 coin

Christianity
Almost 280 million Christians live in the USA, more than in any other country

Missouri river
The longest river in North America (4,086 km). It runs through 7 states

Summer Olympics
The US has won 2,521 Olympic medals (1,022 gold), more than twice as many as the Soviet Union (2nd). The country has hosted the games 4 times

US Route 20

Pacific Railway

UNITED STATES OF AMERICA

Economy
Gross domestic product (GDP) of the US accounts for 25% of the world's total GDP

Largest companies
20 of the world's 50 largest companies by revenue have headquarters in the US

Billionaires
There are 540 billionaires in the US

Inequality
The 400 richest Americans had a combined net worth of $2 trillion in 2013, more than the total net worth of the bottom 50%

Health system
The US has the highest health expenditure per capita

Social support
Around 1/3 of Americans are enrolled in at least one welfare program

Mexican border wall
In 2017, President Donald Trump signed an executive order to build the wall but with no allocation of money, despite it being the biggest promise of his campaign

Oil
USA is the largest oil consumer in the world (20%). 47% of its consumption is from imported oil

Coal
USA has the largest coal reserves in the world but is the 2nd largest producer, after China

PACIFIC OCEAN

Mexican migration
The 1st big wave of Mexican migrants started in the 20th century after Japan's 1907 emigration limitation and the drying up of cheaper Asian labour

The need for Mexican labour increased when the US entered WWI

The Mexican and US governments show no real interest in cooperation concerning the migration issue

0   150   300   600 km

0   100   200   400 miles

American flag
Current flag was designed by Robert G. Heft in 1958 as a school project for which he got an "A". It became the official national flag in 1960

US dollar
US official unit of currency since 1792

The $1 bill represents 45% of total bill production and it "lives" for an average 5.8 years before it is destroyed. In comparison, the $100 bill circulates for 15 years.

It costs 4.9 cents to produce a $1 bill

Up to 90% of $ bills may be contaminated with cocaine

Lake Superior

Cars
There are 797 cars per 1,000 people in the US, 3rd highest ratio in the world, after San Marino and Monaco

Lake Michigan

Lake Huron

Great Lakes
Around 1/5 of the world's fresh surface water is contained in the lake system

Lake Ontario

Lake Erie

US Route 20

Appalachian Trail

BOSTON

Mississippi

CHICAGO

COUNCIL BLUFFS

Route 66

National Road

CUMBERLAND

Appalachian Mountains

WASHINGTON D.C.

Lewis & Clark Expedition

Missouri

VANDALIA

ST LOUIS

National Road
Country's first major highway, constructed between 1811 and 1837

Route 66

Appalachian Trail
World's longest continuous hiking-only trail, about 3,500 km long

Agriculture
Around 8% of the world's agricultural land is within the USA

Appalachian Mountains
The oldest mountains in North America. Their name originated from the Apalachee Indians

Route 66
Finished in 1926, it was a main route for migration to the West. 3,945 km long, it runs through 8 states

Sultana Disaster
This steamboat exploded in 1865 killing up to 1,800 people. It was the greatest maritime disaster in US history, but has long been overshadowed by President Lincoln's assassination the previous day

Meat production
The largest producer of beef and chicken

ATLANTIC OCEAN

Federal lands
28% of all US land is owned by the federal government

Mississippi river
The largest river in North America by discharge (16,792 m³/s) It runs through 9 states (3,765 km)

Prohibition (1920-33)
The ban on the production and sale of alcohol did not decrease consumption. Instead, it encouraged the black market operated by the mafia. Crime increased by 24%

Military budget
World's largest military budget, $824.7 billion (2017-18)

Firearms
There are 88.8 private guns per 100 people in the US, the highest in the world

Mississippi

Racial makeup of the US population

Other 3.1%
Asian 4.7%
Black or African American 12.2%
White 63.7%
Hispanic or Latino 16.3%

Aircraft carriers
The US has 10 operating aircraft carriers, more than any other country

GULF OF MEXICO

Ancestries of the US population

German 15.2 %
Other 28.2%
Irish 10.8%
Native American 2.8%
French 3.0%
African 8.8%
Polish 3.2%
Italian 5.6%
English 8.7%
Mexican 6.5%
American 7.2%

Medicine Lake ★
Every year, up to 10,000 pelicans migrate here from the Gulf of Mexico to breed

Grizzly bear
The largest grizzly bear population in the contiguous US

Glacier National Park
There were 150 glaciers in the park in the mid-19th century. By 2010 there were only 25 active glaciers

★ Roe river
The shortest river in the continent, just 61 m long

Montana

Name
The name of the state comes from the Spanish word "montaña" meaning mountain

Yellowstone National Park
First national park in the US (1872). Home to 75% of the world's geysers including the famous Old Faithful

Black Thunder Coal Mine ★
US's 2nd most productive mine, responsible for 8% of the country's total coal production

Straight borders
Wyoming, Colorado and Utah are the only US states that have only longitude and latitude for boundaries

Voting rights
The first state where women could vote

Wyoming

Population and Area
The lowest population of all the US states (586,107). similar to Luxembourg's. At the same time it is about the size of the UK

Red Desert
Rich in oil, natural gas, uranium and coal 84% of it has been industrialized for mining

CHEYENNE ⊛

"The highest state"
Colorado has the highest mean elevation of all states (2,070 m)

Highest low point
The highest low point (1,011 m) of all 50 states

⊛ DENVER
The city claims to be the place where the cheeseburger was invented

◇ COLORADO SPRINGS
The US Air Force Academy is located here

Colorado

Colorado

Grand Mesa ▲
World's largest flat-topped mountain (3,454 m) with an

◇ PUEBLO
The only city in America to have

★ HELENA

Native Americans
Over 6.5% of Montana's population has Native American heritage. 4th highest in the US

Hells Canyon ★
The deepest river gorge in the US (2,436 m)

Idaho

Invented name
Lobbyist George M. Willing suggested the name "Idaho", stating that it meant "gem of the mountains" in the Shoshone language. In fact, it was made up

⊛ BOISE

Birds of Prey National Conservation Area
The densest population of birds of prey in the world

USA

⊛ SALT LAKE CITY
Until 1868 it was named Great Salt Lake City. Today it is the only state capital with a name with three words

Salt Lake
The state was named after the Sierra Nevada mountain range

Mormons
Around 60% of Utah's are Mormons

Utah

Pando
This 80,000-year-old clonal colony of trees is the heaviest known single organism (6,000 tonnes)

Monument Valley
Great sandstone buttes rising suddenly from the plain

"Going to the Sun Road"
This road, located in Glacier National Park, is considered one of the most beautiful drives in the world

"Gem State"
The official nickname of Idaho

Cloudiest in the US
Seattle is at least partially covered by cloud 6 out of 7 days a week between April and October

★ Dry Falls
Today - almost dried out. During the last Ice Age - the largest waterfall on Earth with a flow 10 times greater than the flow of all today's rivers

SEATTLE
Jimi Hendrix was born here and Ray, Charles and Quincy Jones started their careers here

OLYMPIA ⊛

Washington

BEAVERTON
Headquarters of Nike is here

Two-sided flag
The only state that has a flag with two different sides

⊛ SALEM

Cheese factory
The Tillamook factory is one of the world's largest cheese factories. Their cheddar is still made based on a 100-year-old recipe

Oregon

Honey fungus ★
The largest fungus in the world is located here. It covers 8.9 km² and is 2,400 years old. It thrives thanks to low competition for soil and nutrients in the Malheur National Forest

Gambling
Gamblers lost $92.27 billion in the US (2007)

Name
The state was named after the Sierra Nevada

Gold
The US is the 4th largest gold producer in the world (around 5% of global production)

Crater Lake ★
The deepest in the US and 10th in the world (594 m)

Population growth
In 1930, Nevada had the lowest population of 91,058 people. It rapidly increased after 1931 when gambling was legalized here

Jack Johnson became the first African-American world heavyweight boxing champion in 1908. His defeat of James Jeffries in Reno in 1910 was dubbed "the fight of the century"

⊛ RENO

Nevada
The "Silver State"

Prostitution
Nevada is the only state where prostitution is legal

California Gold Rush
The first gold found here in 1848 attracted around 300,000 people to California

Population
The most populous state in the United States with 38 million people - comparable to Poland

Folsom State Prison
One of the first maximum security prisons in the country. Famous thanks to Johnny Cash's concerts performed inside

⊛ CARSON CITY

Nevada Test Site
928 nuclear tests have been

Yosemite National Park
Park famous for its beautiful scenery, waterfalls,

⊛ SACRAMENTO

SAN FRANCISCO ⊛

State Route 1

Mavericks
Famous surfing location with waves up to 8 m high. Gained its popularity as recently as 1990

State Route 1
One of the most beautiful coastal routes in the world

California's economy
The largest economy in the US. If it was a separate country, it would be the 6th richest in the world, behind the UK and ahead of France

Sequoia National Park
Home to the largest living tree, General Sherman. Its trunk is 31 m in circumference

California
"Eureka"
The state motto, referring to the discovery of gold

San Bernardino County
The largest county in the US, slightly smaller than the state of West Virginia

Copper
Arizona is the largest copper producer in the US

LAKE HAVASU CITY
One of the bridges in this city was bought from London, UK for £2.5 million and shipped here in 1968

FALLBROOK
Known as the "Avocado Capital of the World"

Busiest border crossing
Busiest border crossing in the Western Hemisphere. In 2015, around 14.4 million cars crossed the San Ysidro Port of Entry

Indian lands
In Arizona, Indian lands occupy the largest percentage of land, out of all US states

Saguaro cactus blossom
The national flower of the state and the largest cactus in the country

LAS VEGAS
LOS ANGELES
SAN DIEGO
PHOENIX

Arizona

Grand Canyon National Park
The canyon is up to 29 km wide and 1,857 m deep. Colorado river began to carve the rocks of the Colorado Plateau 5–6 million years ago

Four Corners
The only place in the US where 4 states meet

New Mexico

SANTA FE
The highest (2,194 m) and the oldest (1610) state capital in the US

White Sands Missile Range
The first atomic bomb was detonated here on 16 July 1945

No water
Only 0.2% of its area is covered with water, the lowest of all states

"Latino state"
New Mexico has the highest percentage of Latinos and Hispanics

US territorial expansion 1783–1853

1783
Original thirteen states ceded by Great Britain after the Treaty of Paris

1803
Louisiana Purchase from France

1818–21
Red River acquired from Great Britain. Florida ceded from Spain

1845–6
Texas annexation. Oregon Territory gained from Great Britain

1848–53
Mexican Cession after the Mexican-American War. Gadsden Purchase from Mexico

0   75   150   300 km
0   50   100   200 miles

PACIFIC OCEAN

KAUA'I
O'AHU
HONOLULU
MOLOKA'I
MAUI
HAWAI'I

Hawaii
Admitted in 1959, the most recent, 50th state

Mauna Kea (4,205 m)
From its oceanic base (-6,000 m) to its peak, it's the tallest mountain on Earth. Considered the best astronomical observation station on the planet (dry climate, high altitude, low light pollution, 13 telescopes)

Ka Lae
The southernmost point of the 50 US states, lying at a similar latitude to the capital of Jamaica, Kingston

Pineapple
Over one-third of the world's pineapple supply comes from these islands

Widest state
The islands of Hawaii spread over 2,450 km

Lake Superior

Lake Michigan

CHICAGO ☆

Colonization
French fur traders were the first Europeans to reach this area in the 17th century. During the 19th century, emigrants from Germany and Scandinavia also settled here

"America's Dairyland"
Immigrants' cheese-making traditions combined with convenient geography gave the state its reputation as a leader in dairy production

Wisconsin

Summerfest
Located in Milwaukee, it is the world's largest music festival: 30 ha, 11 stages, over 1,000 performances, lasts 11 days and attracts up to 900,000 people

Harley Davidson
Founded in Milwaukee in 1903, it was one of 2 American motorcycle manufacturers that survived the Great Depression

MILWAUKEE ○

MADISON ○
In 1829, a former federal judge bought 4 km² of land and named the soon-to-be city after President Madison

McDonald's
First McDonald's franchise was opened in Des Plaines in 1955

Home Insurance Building
The first skyscraper (42 m). It was the highest building in the world from 1884 to 1889

Illinois

☆ SPRINGFIELD

Slavery abolition
The first state to ratify the Thirteenth Amendment that abolished slavery

The tallest man
Robert Wadlow was 2.72 m tall. He died in 1940 when he was 22

Grey wolf
The largest population of grey wolves outside of Alaska

Minnesota

"Pig's Eye"
Original name of the state's capital, after a French-Canadian fur trapper: Pierre "Pig's Eye" Parrant

MINNEAPOLIS ○ ○ ST PAUL

Water skiing
First water ski ride at Lake City in 1922

University of Minnesota
☆ Country's first open-heart surgery
☆ Country's first bone marrow transplant
☆ World's first heart operation with the use of a deep freezing technique

Fort Atkinson
The only fort built to keep peace between various Native American tribes

Cornell College
One of only a few colleges that offer OCAAT – "one course at a time" studying system

Iowa

DES MOINES ○
Named after the Des Moines river, "River of the Monks" in French. It is a hub of the US insurance industry

Water borders
The only US state whose east and west borders are totally covered with water (Missouri and Mississippi rivers)

Most neighbourly states
Missouri and Tennessee are bordered by 8 states, more than any other state

HANNIBAL
The story of "The Adventures of Tom Sawyer" took place in the fictional town of St Petersburg inspired by Hannibal where Mark Twain was raised

Sunflowers
North Dakota grows around half of the US's sunflowers. It is estimated that blackbirds eat up to $10 million worth of them every year

Male population
2nd highest male population (51.25%), after Alaska

North Dakota

BISMARCK ○
The city was renamed by the Northern Pacific Railway in 1873, in Otto von Bismarck's honour, and to attract German investors and settlers

FARGO ○
This Academy Award-winning movie was named after the city but none of it was shot at this location

Black-footed ferret
North America's most endangered mammal has been reintroduced in this state

Lack of trees
Only 1.8% of the state's area is covered by trees

LINCOLN ○

Grizzly bear
The first expedition that crossed the US encountered its first grizzly bears here. They had disappeared from these lands by 1900

Oahe Dam
As a result of its construction, 340 km² of Indian reservations were lost with a dramatic effect on native inhabitants

South Dakota

PIERRE ○
2nd least-populous state capital after Montpelier, Vermont

Ashfall Fossil Beds ☆
Site with well-preserved fossils of mammals from the Miocene geological epoch which died from a Yellowstone eruption 10–12 million years ago

USA centre ☆
Geographical centre of the contiguous states

Nebraska

Bleeding Kansas
Violent political confrontations between anti- and pro-slavery forces from 1854 to 1861. Eventually, Kansas entered the Union as a free state

TOPEKA ○ ○ KANSAS CITY

Kansas

Mount Rushmore National Memorial
18 m tall. Construction took 14 years and was completed in 1941. The idea was to promote tourism in the region. It features the heads of 4 presidents: Washington, Jefferson, T. Roosevelt and Lincoln

Prairie
The state is mostly covered by plains

Naval Ammunition Depot ☆
The largest US ammunition plant during World War II. It was located at Hastings. It provided 40% of ammunition for this war

Cheyenne Bottoms
The largest wetland in the interior of the US. This relatively small area is a critical staging post for millions of migrating birds

Rivers
Nebraska has the greatest length of rivers of all states

Legislature
The only US state to have a unicameral legislature. It is also unique because it is nonpartisan

JEFFERSON CITY ○

Missouri

ST LOUIS ○ ○ ○
Iced tea was popularized here at the World's Fair of 1904

Midwest
One of four official geographic regions of the US. It consists of: Illinois, Indiana, Iowa, Kansas, Michigan, Minnesota, Missouri, Nebraska, North Dakota, Ohio, South Dakota and Wisconsin

First female mayor
The first female mayor in the US was Susanna Madora Salter, elected in Argonia in 1887

First Pizza Hut
First Pizza Hut was opened in Wichita in 1958

Hot & cold
Warsaw holds the state record for both the lowest (-40 °C) and highest (48 °C) temperatures

USA

First Walmart
Opened in Rogers in 1962. Today, it employs 2.3 million people worldwide and has more than 11,000 stores in 27 countries

Oklahoma City bombing
The Alfred P. Murrah Federal Building bombing in 1995 was the deadliest terrorist attack on US soil until 9/11. 168 people died by the hands of anti-government militants

Dust Bowl
Period during the 1930s when dust storms damaged American and Canadian prairies

Oklahoma

OKLAHOMA CITY ○

Trail of Tears
Forced removals of Native Americans from their lands to designated territories west of the Mississippi river. Over 4,000 died of exposure, disease or starvation

Execution
Texas was the 1st state to carry out an execution by lethal injection, in 1982

Capital punishment
Around 1/3 of the country's executions have been performed in this conservative state

Size
2nd largest state in the US, after Alaska. Its area is similar in size to Afghanistan and it's larger than any country in contiguous Europe

Population
2nd largest population (27.9 million) of all US states, after California

○ DALLAS
President J. F. Kennedy was shot here on 22 November 1963. Around half of Americans believe in a conspiracy theory around this event

"Six flags over Texas"
Slogan used to describe 6 countries that once ruled these lands: France, Spain, Mexico, Republic of Texas, the Confederate States of America and the United States of America

Arkansas

LITTLE ROCK ○
City's name comes from a rock formation on the local river, named in French "le petit rocher"

The Crater of Diamonds Mine is the only active diamond mine in the US

Hattie Caraway
1st woman to be elected as a US Senator, in 1932

Lone Star Republic
Texas was an independent state between 1836 and 1846 before annexation by the US

Prairie Chapel Ranch
Famous ranch owned by the ex-president, George W. Bush

Johnson Space Center
One of the ten major NASA field centres. The first words from the Moon were received here, spoken by Neil Armstrong: "Houston, Tranquility Base here. The Eagle has landed."

Texas

AUSTIN ○

○ SAN ANTONIO

HOUSTON ○

Midland Basin Wolfcamp shale
Possibly the largest reserves of crude oil ever discovered in the US, rich in 20 billion barrels of oil

Edwards Aquifer
One of the world's richest artesian aquifers and a source of drinking water for over 2 million people

Oil
Texas has the highest reserves of crude oil in the US

Texas blind salamander
lives in the underwater caves of Edwards Aquifer. It has developed external gills through which it absorbs oxygen from water. The salamander has no eyes or skin pigment

Desert
Texas is associated with a dry desert landscape but only 10% of its surface is desert

Big Bend National Park
Contains over 1,200 species of plants, 450 species of birds, 75 species of mammals and 56 species of reptiles

Mississippi

Elvis Presley was born here in 1935

TUPELO ○

Civil War
Around 59,000 out of 78,000 Mississippian soldiers who entered the Confederate army were killed

○ JACKSON
"The City with Soul"

Isaac Ross
In 1836, he freed his slaves in his will. They travelled to Africa where they became one of the founders of Liberia

First African-American officers
African-American officers fought in the Confederate Louisiana Native Guards in 1862

Louisiana
Named after the French king Louis XIV

Hurricane Katrina
The costliest tropical storm, causing over $100 billion of damage. 1,836 people died

BATON ROUGE ○

NEW ORLEANS ○
Jazz originated here in the late 19th century

Louisiana Purchase
The US paid France $15 million for the Louisiana Territory in 1803. These lands extended over 2.14 million km² west of the Mississippi river

Tiber Oil Field
This oil field is owned by British Petroleum. It required the drilling of a well 11,939 m below sea level, under 1,259 m of water, and is one of the world's deepest wells

GULF OF MEXICO

0   75   150   300 km
0   50   100   200 miles

Lake Superior
The largest wholly-freshwater lake in the world by area, similar in size to Austria

Lake Michigan–Huron
Hydrologically, it is a single lake, as waters of both basins, flowing through an 8-km-wide strait, are in near-equilibrium. Counting Michigan and Huron as one lake makes it the world's 2nd largest freshwater lake by area

MANITOULIN ISLAND
World's largest island in an inland body of water. Its lake, Lake Manitou, is the world's largest lake on a freshwater island

Lake Ontario
The smallest of the Great Lakes. It is similar in size to Lake Erie but holds about 4 times the water volume because it is much deeper

Ivy League
- Brown University, RI
- Columbia University, NY
- Cornell University, NY
- Dartmouth College, NH
- Harvard University, MA
- University of Pennsylvania, PA
- Princeton University, NJ
- Yale University, CT

Smallest capital
Montpelier, Vermont has the lowest population of all US state capitals (7,855 people in 2010) and is the only state capital without a McDonald's

Russo-Japanese War
Russo-Japanese War (1904–5) ended with a treaty signed in Portsmouth, New Hampshire

New Hampshire primary
Every four years it is the first nationwide party primary election

Crime
Vermont has the lowest reported violent crime rate of all US states (99 per 100,000 people)

Lobster
Of all the lobster caught in the north Atlantic Ocean, around half is caught off the coast of Maine

Harvard University
First university in the US (1636)

Team sports
Basketball was invented in Massachusetts in 1891 and volleyball in 1895

Prohibition
Only Connecticut and Rhode Island did not ratify the 18th Amendment introducing prohibition

Infrastructure
New Jersey's railway and highway systems are the densest in the US

Maine
The only US state that borders only one other state

Only state in the US whose name is one syllable

Rhode Island
The smallest US state (4,002 km²)

USS Nautilus
World's first nuclear-powered submarine was launched in 1954 in Groton

New Jersey

Population density
New Jersey has the highest population density of all US states, 467.2 people/km². Around 90% of the population live in urban areas

NYC Subway
The highest number of stations (472) in the world

Delaware
Named after English nobleman Thomas West, 3rd Baron De La Warr

Maryland

National Anthem
Written by lawyer Francis Scott Key, inspired by the bombardment of Fort McHenry

District of Columbia

The Pentagon
It's actually located in the state of Virginia, on the outskirts of Washington D.C.

Constitution
Delaware was the first state to ratify the constitution in 1787

"The Virgin Queen"
The state of Virginia was named after "The Virgin Queen" Elizabeth I of England

Battles of Saratoga
Turning point of the American Revolutionary War. This victory convinced the French to become allies, changing the balance of forces

New York

West Point Academy
Every year, around 1,000 out of 1,300 cadets graduate

Rockville Bridge
World's longest stone arch railway viaduct (1,164 m)

Pay equality
New York is the state closest to reaching pay equality

Pennsylvania

Battle of Gettysburg
Civil War's turning point (1863) with the highest number of casualties in the entire war

Battle of Antietam
The bloodiest battle (1862) on American soil with 22,717 casualties and losses

Mother's Day
First celebrated in West Virginia in 1908

West Virginia

Virginia

8 US presidents were born in Virginia: W. H. Harrison, Jefferson, Madison, Monroe, Taylor, Tyler, Washington and Wilson

Lake Ontario

Lake Erie
The Iroquoian word for "long tail"

Lake Erie

First police car
Operating from 1899, it was a wagon run by electricity that could drive for 48 km at a speed of 26 km/h before its battery had to be recharged

First ambulance
1st hospital-based ambulance was established in Cincinnati in 1865

Ohio

Neil Armstrong
The first person to walk on the Moon was born in Ohio

Lake Huron

Boats
The 2nd highest number of registered boats in the US

Shoreline
The longest freshwater shoreline in the world

First tunnel
First international underwater railway tunnel was opened here in 1891

Michigan

Deforestation
Before the arrival of Europeans, over 80% of Indiana was covered by forest. Today, it is only 17%

Tri-State Tornado
Deadliest tornado in the history of the US (1925). Killed 695 people in Missouri, Illinois and Indiana

Indiana

The country's first long-distance auto race took place here in 1911

The Ford Model T was produced in Detroit and sold 16.5 million units. In 1914, 90% of the world's cars were Fords

Lake Michigan
The only one of the Great Lakes that is located entirely in one country – USA

First flight
In 1903, the Wright brothers made their first controlled and sustained flight in a powered plane from here

Cape Lookout National Seashore
Three barrier islands (90 km long) covered with beautiful and remote beaches

ATLANTIC OCEAN

American Civil War 1861-5
The war was fought between the Union (23 slave-free states and 5 "border" slave states that did not secede from the Union) and the Confederacy (11 southern secessionist slave states)

☐ Union states
☐ Confederate states

Battle of Fort Sumter
The Civil War started here on 12 April 1861 with the Confederate bombardment of Fort Sumter

USA

● RALEIGH

North Carolina

Carolina Gold Rush
The first gold rush in the US started here in 1799 after Conrad Reed, a 12-year-old boy, found a 7-kg gold nugget

Jack Daniel's
This top selling American whiskey was founded in Lynchburg in 1875

Mammoth Cave
World's longest cave system (652 km)

Abraham Lincoln, the President of the Union, and Jefferson Davis, President of the Confederacy, were both born in Kentucky

Great Smoky Mountains National Park
With 13.3 million visitors in 2016, it is the most visited national park in the US

NASHVILLE ●

Tennessee

★

Marshall Space Flight Center
The largest NASA centre, responsible for spacecraft propulsion research

Alabama

Alabama Fever
First great American land boom that started in 1817 with settlers and their slaves claiming the lands ceded by Native Americans

MONTGOMERY ●
Chosen as the first capital of the Confederate States of America (1861)

Montgomery bus boycott
In 1955, Rosa Parks was arrested for refusing to give up her seat to a white person

Excerpt of French Louisiana

MOBILE ●

German torpedo
In 1942, the Robert E. Lee steamboat was torpedoed and sunk by a German submarine. 15 passengers died, all of whom were survivors of previous German torpedo attacks

Deepwater Horizon oil spill
The largest marine oil spill in history (4.9 million barrels) occurred in 2010. The oil leaked for 5 months

Oil spill area

CHARLOTTE ○

First library
South Caroliniana library, in Columbia, was the 1st (1840) free-standing academic library building in the US

● COLUMBIA

South Carolina

Marine Corps Recruit Depot, Parris Island
US Marine Corps training facilities, famous for training scenes from Stanley Kubrick's movie Full Metal Jacket

Country's 2nd largest financial centre after New York City

● ATLANTA
World's busiest airport with around 100 million passengers annually

Berry College
The largest contiguous university campus in the world (110 km²)

Coca-Cola
Invented in 1886 in Atlanta

Georgia

King George II
The state was named after this king of Great Britain and Ireland

Okefenokee Swamp
The largest "blackwater" swamp (acidic, darkly stained) on the continent

TALLAHASSEE ●

First scheduled airline flight
Took place in 1914. The ticket for a 23-minute flight was $400

Port of Jacksonville
During the Gulf War (1990-1) it was the US's busiest military port, responsible for supplies for the operation

Cape Canaveral
US spacecraft launching site, formerly named Cape Kennedy. The first rocket was launched in 1950

Tornadoes
Florida is the state with the highest frequency of tornadoes, 12.2 per 16,000 km² annually

Florida

TAMPA ★

ST PETERSBURG ●

Abandoned wells
There are over 27,000 abandoned oil and gas wells under the waters of the Gulf

Everglades National Park
One of only 3 places in the world declared a Wetland of International Importance, World Heritage site and International Biosphere Reserve

Known as the "Venice of America" due to its extensive network of canals (266 km)
FORT LAUDERDALE

GULF OF MEXICO

0   75   150   300 km
0   50   100   200 miles

Mexico-US barrier

The 3,145-km-long continental border is protected by a series of walls, fences, sensors and cameras. 463,000 Border Patrol apprehensions were made in 2010

Rio Grande

Rio Grande

In 2001, the river failed to reach the Gulf of Mexico because of a 100-m-wide sandbar that formed at its mouth

"The volleyball wall"

Every year, residents of Naco, Arizona, join residents of Naco, Mexico, for a match at the fence that separates USA and Mexico.

Birth control pill

Saponin, found in inedible Mexican yams, was a key ingredient in the development of the combined oral contraceptive pill

Biodiversity

World's 4th most biodiverse country. Over 200,000 different species live here

Avocado

World's largest avocado producer, supplying more than 30% of the world's market

Diverse climate

Mexico has one of the most diverse climates, ranging from tundra to Mediterranean and tropical climates

Avocado cultivation

In 2013, Mexico's land area dedicated to avocado production was 1,680 km². This corresponds to the surface area of this square

GUADALUPE

At some point in the 19th century, this little island was inhabited by as many as 100,000 goats

Baja California

GULF OF CALIFORNIA

Sierra Madre Occidental

Reptiles

The 2nd most biodiverse country in reptiles (after Australia) with 804 species

Silver

Leader in silver production (5,400 tonnes in 2013)

The World's aquarium

The waters around Baja California and the Gulf of California form a habitat for about 1/3 of all marine mammal species on the planet

MEX

Transoceanic countries

Mexico is one of 22 contiguous countries that border 2 or more oceans

Gulf of California

Classified as a sea

Vaquita

This porpoise, endemic to the Gulf of California, is the world's most endangered cetacean

American Cordillera

Almost continuous chain of mountain ranges that stretches from Alaska to Antarctica

MARÍAS ISLANDS

Penal colony with a federal prison

GUADALAJARA

Lake Chapala

This main source of water for Guadalajara is in danger. Deforestation around its coast led to soil erosion which increased the sedimentation of the water. This resulted in a decrease in the lake's depth which raised its temperature. In turn this led to increased evaporation

REVILLAGIGEDO ISLANDS

Revillagigedo Islands

Never inhabited until their discovery in 1533, home to many endemic species. Colloquially called Mexico's "little Galapagos"

Longest sea survival

Longest sea survival

In 2012, two men went fishing, got caught in a storm and drifted 10,700 km for 438 days. They lived off fish and birds and drank turtles' blood. Only one of them survived

PACIFIC OCEAN

0	100	200	400 km

0	50	100	200 miles

Mexico in 1824

Mexico today

Cinco de Mayo

5 May is a holiday in Mexico, celebrating Mexico's victory over the French in the Battle of Puebla of 1862. In the US, this day has taken on a completely different meaning and is a symbol of friendship between these nations

Monarch butterfly migration

They migrate from Canada to Mexico and back. None of them complete the trip: they stop, lay eggs and newly emerged butterflies continue the journey. It takes 5 generations to complete the migration

Rio Grande

Native foods of Mexico

Avocado, beans, chilli, chocolate, tomato, vanilla, courgette and many more

"Dead Zone"

Area where marine life cannot be supported at times due to low oxygen levels. It is caused by runoff dumped by the Mississippi river

CO_2

Gulf Stream

One of the strongest warm currents in the world, influencing the climate of North America and Europe. It originates in these warm waters

GULF OF MEXICO

Ninth largest body of water in the world. First explored by Amerigo Vespucci in 1497

Gulf Stream

CO

Tomatoes

Originated in South America, but first widely eaten in Mexico. Native tomatoes were yellow and small

"Mexican miracle"

From the 1940s until the 1970s, Mexico's economy grew at an incredible rate of 3 to 4% annually

National University of Mexico

The oldest university in North America (1551) is in Mexico City

Chicxulub Crater

The asteroid that hit this point 65 million years ago, probably caused the extinction of the dinosaurs

CANCÚN

Great Pyramid of Tenochtitlan

In 1487, the Aztecs sacrificed up to 80,000 prisoners here during the ceremony of re-consecration

Pico de Orizaba (5,610 m)

Highest mountain in Mexio and 3rd highest in North America

Yucatán peninsula

The first European encounter with an advanced civilization in the Americas (in 1517) took place here

Tourism

Today, the most popular tourist destination in Mexico. In 1970, the area had only three residents

MEXICO CITY •

7th largest metropolitan area in the world and the largest in North America with 21 million people

VERACRUZ

Guerrero

The lowest ranked state in the 2015 Mexico Peace Index

Great Pyramid of Cholula

The world's largest pyramid by volume

First railway

Route of the first railway line in Mexico, opened in 1873

Cenotes

The Yukatán peninsula has thousands of these vertical, water-filled sinkholes. Cenotes were once used by the ancient Maya for sacrificial offerings. Today, they offer some of the most "magical" and dangerous dives on the planet

Mexican Drug War

It has taken the lives of over 60,000 people. It is estimated that Mexican drug cartels may have up to 100,000 members

Sacrifice

The Aztecs sacrificed 1% of their population every year, approximately 250,000 people

Independence
When Belize became independent in 1981, Guatemala refused to recognize the new country as it considered itself the rightful inheritor

Maya Biosphere Reserve
The largest tropical forest in Central America. Home to jaguars, pumas, spider monkeys, crocodiles and freshwater turtles among others

Ban on fishing
In 2010, Belize was the first country in the world to ban bottom trawling in order to protect its reef

Great Blue Hole
Giant (300 m wide, 108 m deep) sinkhole. Before the sea level rose around 60,000 years ago, it was a cave in the ground

HIV
Around 60% of Central America's HIV infections are reported in Honduras

BELMOPAN

BELIZE
From 1862 to 1981, it was called British Honduras

Belize Barrier Reef
Second-largest barrier reef in the world (300 km long)

Volcanoes
6 of Guatemala's 29 volcanoes have erupted in the recent past

GUATEMALA

Mayan civilization
Guatemala is the hub of Mayan civilization, famous for architecture, advanced mathematics and astronomy. Half of the population are direct descendants of the Mayans

Río Plátano Biosphere Reserve
Beautiful rainforest, rich in wildlife. Due to cattle grazing, hunting and deforestation, it was placed on UNESCO's World Heritage in Danger list, one of 54 in the world

Murder rate
The highest murder rate in the world (90 murders per 100,000 people)

Rain of Fish
Once a year in the town of Yoro, fish get caught in ocean waterspouts and then fall from the sky from rainstorms

Poverty
One of the poorest states in this hemisphere

HONDURAS

• GUATEMALA CITY

Monja blanca
National flower of Guatemala and a white symbol of peace. This orchid flowers only after 15 years of life

TEGUCIGALPA
Its airport is the 2nd most dangerous in the world

MMA
Mixed martial arts is one of the most popular sports in Guatemala

△ SAN SALVADOR

Bosawás Biosphere Reserve
Largely unexplored, 2nd largest rainforest of the Western Hemisphere

EL SALVADOR

NICARAGU

Largest country in Central America. Similar in size to Greece

"The Lighthouse of the Pacific"
The Izalco volcano erupted regularly for two centuries

Gulf of Fonseca
Beautiful coasts with forests, sandy beaches, rocky cliffs and islands. Despite past dense settlement, the environment has had relatively little modification

Lake Managua

MANAGUA

El Salvador is the only Central American country with no access to the Caribbean Sea

Bull shark
Known for living in the oceans but also enters fresh water through rivers

Lake Nicaragua

Lake Nicaragua
Even though it is a freshwater lake, sharks, sawfish and tarpons all live here

Ring of Fire

Ring of Fire
Oceanic ridges and trenches, 40,000 km long, encircling the Pacific Ocean

Most of the world's earthquakes take place within its boundaries

Almost 75% of all active volcanoes are located here

Golden toad
This toad was last seen in 1989. Its extinction was the first to be blamed on human-made global warming

Renewable energy
Costa Rica supplies 98% of its energy demand from renewables

PACIFIC OCEAN

"Sustainable country"
Costa Rica is the most stable and progressive country in the region

Identified by the New Economics Foundation as the greenest country in the world in 2009

Costa Rica plans to become the first carbon-neutral economy by 2021

Panama Canal

Atlantic Ocean — Gatun Lake — Pacific Ocean
↕ 26 m

• Length: 77 km
• Crossing time: 8 hours
• Opened: 1914
• Cost: $375 million
• Deaths: 22,000 workers
• Traffic: 12,000–15,000 vessels per year

PANAMA CITY

Gatun Lake

Gatun Lake
Artificial lake made for this project. Loses 202,000 m³ with every ship that crosses the canal

★ locks

| 0 | 30 km |
| 0 | 20 miles |

| 0 | 50 | 100 | 200 km |
| 0 | 30 | 60 | 120 miles |

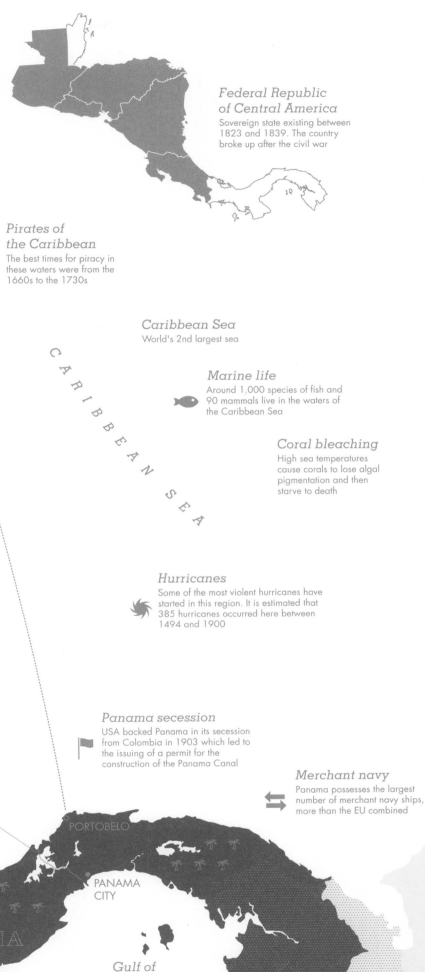

WAN ISLANDS (Honduras)

cept for the Honduran navy garrison,
e islands are uninhabited. Radio Swan
s broadcasting during the Bay of Pigs
asion of Cuba. The transmitter was
er removed and transferred for use in
e Vietnam War

"Coral economy"

It is estimated that fishing and
diving connected to coral reefs
provide Caribbean countries
with annual revenues between
$3.1–$4.6 billion

Pirates of the Caribbean

The best times for piracy in
these waters were from the
1660s to the 1730s

Federal Republic of Central America

Sovereign state existing between
1823 and 1839. The country
broke up after the civil war

Havana - Portobelo trade route

Piracy in the
Caribbean Sea has
thrived thanks to the
attacks on this trade
route

Havana - Portobelo trade route

Caribbean Sea

World's 2nd largest sea

C A R I B B E A N   S E A

Marine life

Around 1,000 species of fish and
90 mammals live in the waters of
the Caribbean Sea

Banana Wars

Numerous (1898–1934) US military
actions in Central America and the
Caribbean, aimed at securing
American interests in the south

Coral bleaching

High sea temperatures
cause corals to lose algal
pigmentation and then
starve to death

Ometepe

The island in Lake Nicaragua is the
highest rising island in a lake in
the world, with a perfectly shaped
volcanic peak, Concepción, at
1,610 m above sea level

Hurricanes

Some of the most violent hurricanes have
started in this region. It is estimated that
385 hurricanes occurred here between
1494 and 1900

Panama disease

Devastating plant disease
which, during the 1950s,
destroyed most banana crops
in Central America

Panama secession

USA backed Panama in its secession
from Colombia in 1903 which led to
the issuing of a permit for the
construction of the Panama Canal

Merchant navy

Panama possesses the largest
number of merchant navy ships,
more than the EU combined

Panama Canal

Tolls collected here drive 1/3 of the
economy. The canal was one of the most
difficult constructions ever undertaken,
mainly due to yellow fever and malaria
which infected and killed workers. In 1884,
the death rate was over 200 per month

COSTA RICA

SAN JOSÉ
Named after Joseph of
Nazareth, it was established in
1738 and is one of the safest
cities of Central America

PORTOBELO

PANAMA CITY

Mosquito Gulf

Costa Rica
means "Rich Coast"

PANAMA

Gulf of Panama

Rainforest

Rainforest covers around
58% of the total land
area of Panama

El Diquís Dam

Currently in the planning
phase, it will be the biggest
hydroelectric dam in Central
America and will meet 1/3 of
the country's energy demand

Darién Gap

Inaccessible forests and
swamplands with no roads. It is the only
missing link of the Pan-American Highway.
It is occupied by the Revolutionary Armed
Forces of Colombia responsible for
drug trafficking and kidnappings

Volcán Barú (3,475 m)

The only place in the world where
you can see the sun rise on the
Atlantic and set on the Pacific

ATLANTIC

Boxing
With 37 gold medals, Cuba is the 2nd most successful country in Olympics boxing (USA has 50 gold medals)

CASTAWAY CAY ★
Private island used exclusively as the Disney Cruise Line ships' port

Freedom
In 2016, according to the State of World Liberty Index, people of The Bahamas came 6th highest for personal and economic freedoms

THE BAHAMAS

NASSAU

Islands
Over 700 islands are estimated to make up The Bahamas

Travel
Since 2013, Cubans require only a valid passport to leave the country. However, this document costs five months average salary

Diana Nyad
In 2013, at the age of 64, she became the first person to swim 180 km from Cuba to Florida

HAVANA

CUBA

Communism
Cuba has been ruled by the Communist Party of Cuba since 1965. It was not until 2014 that Cuba opened its borders once more to Americans

Bay of Pigs Invasion
1961 unsuccessful invasion of the island undertaken by a paramilitary group made up of Cuban emigrants and exiles. The operation was planned and supported by the CIA

HIV
In 2015, Cuba became the first country to eradicate mother-to-child transmission of HIV

HAITI

Tax haven
One of the most popular offshore financial centres, especially among Americans. Among the world's "tax havens" it has the highest value of assets under management in offshore funds

CAYMAN ISLANDS (U.K.)
The islands were relatively late to be populated with the first permanent recorded resident around 1661

Cayman Trough
★ The deepest point in the Caribbean Sea (7,686 m)

Guantanamo
107 people were detained, out of which 90 were held for more than 10 years. The yearly cost to imprison a detainee is around $3 million

PORT-AU-PRINCE

Churches
Not counting Vatican City, Jamaica has the highest number of churches per km² in the world

JAMAICA
KINGSTON

PORT ROYAL
Main pirate base in the 17th century

Poverty
Haiti is one of the poorest countries in the world. Around 80% of the population live in poverty

2010 earthquake
During the tragic 2010 earthquake, over 300,000 Haitians lost their lives and 1 million were left homeless

Cuban immigrant population in the US

Year	Number of immigrants
1980	608,000
1990	737,000
2000	873,000
2006	936,000
2010	1,105,000
2013	1,144,000

Usain Bolt
Jamaican, considered the best sprinter ever, was the first to hold both the 100 m and 200 m world records. He is also an 8-time Olympic gold medallist and 11-time World Champion

Size
Caribbean 239,681 km² ≈ Ghana 238,537 km²

Population
Caribbean 43,489,000 people ≈ Argentina 43,417,000 people

CARIBBEAN SEA

| 0 | 75 | 150 | 300 km |
| 0 | 50 | 100 | 200 miles |

LESSER ANTILLES

VIRGIN ISLANDS (U.K.)
More than half of the islands' labour force is foreign born

ANGUILLA (U.K.)

St Martin
The smallest island in the world to be shared between two sovereign nations, France and the Netherlands

ST-MARTIN (Fr.)

SINT MAARTEN (Neth.)

ST-BARTHÉLEMY (Fr.)

Taíno language
Hammock, tobacco, barbecue, hurricane – these words, among many others, come from the extinct Caribbean language Taíno

VIRGIN ISLANDS (U.S.A.)
Ruled by the English, Dutch, French, Spanish and Danish. The US bought the islands from Denmark for $25 million in gold

SABA (Neth.)
ST EUSTATIUS (Neth.)

BASSETERRE

ANTIGUA & BARBUDA

ST JOHN'S

Mount Obama (402 m)
The highest point on Antigua was named after US president Barack Obama in 2009

ST KITTS AND NEVIS
The smallest and least populated (56,000 people) country in the Western Hemisphere

MONTSERRAT (U.K.)

GUADELOUPE (Fr.)
Columbus discovered the pineapple here in 1493

Boiling Lake
One of the world's largest thermal lakes located in the jungle of Morne Trois Pitons National Park

Columbus – First voyage

TURKS AND CAICOS ISLANDS (U.K.)

DOMINICA

ROSEAU

"Nature Island"
Dominica's relatively small, mountainous rainforest includes 360 rivers and 9 volcanoes

Mount Pelée (1,397 m)
The deadliest volcanic disaster of the 20th century killed about 30,000 people in 1902

DOMINICAN REPUBLIC
The most visited Caribbean country (5 million tourists per year)

Milwaukee Deep
Deepest place in the Atlantic Ocean (8,605 m) and 10th deepest spot on Earth's seabed

MARTINIQUE (Fr.)
One of the places where the exiled French protestant Huguenots were sent

SANTO DOMINGO

Mountains
Gros Piton (798 m) and Petit Piton (750 m) – these characteristic tropical mountains are symbolized on the country's national flag

CASTRIES

ST LUCIA

Exports
Mainly bananas, beer and petroleum oil

Pico Duarte (3,175 m)
Highest peak in the Caribbean

PUERTO RICO (U.S.A.)

El Yunque National Forest
The only tropical rainforest belonging to the US National Forest System

ST VINCENT AND THE GRENADINES

KINGSTOWN

BARBADOS

BRIDGETOWN

Botanic gardens
Home to one of the oldest botanic gardens in the Western Hemisphere (1765)

Mount Gay Rum
The oldest existing rum brand (1703)

"The bearded ones"
The meaning of the word "Barbados"

ANTILLES
Archipelago divided into two groups: Greater Antilles (Cayman Islands, Cuba, Puerto Rico, Hispaniola and Jamaica) and Lesser Antilles

ST GEORGE'S

GRENADA
Also known as the "Island of Spice"

The name "Grenada"
The island was renamed after the Spanish city of Granada by the sailors who ventured there at the beginning of the 18th century

TOBAGO

Tobacco pipe
The name of the island comes from the Carob word for tobacco pipe (tavaco), which has a similar shape to the island

TRINIDAD

ABC Islands
These three islands are often referred to as the ABC islands

Climate
Unlike the rest of the Caribbean, the ABC islands have a dry climate, similar to the Mediterranean, with low humidity and little rainfall

TRINIDAD AND TOBAGO

PORT OF SPAIN

Biodiversity and asphalt
Flora and fauna on Trinidad is the richest of all the Caribbean islands. The island also contains one of the largest natural sources of asphalt on the planet, Pitch Lake

CURAÇAO (Neth.)
Prostitution is legal here, but only for foreign women and only in one specific brothel

ARUBA (Neth.)

BONAIRE (Neth.)
Famous for large populations of flamingos and donkeys

0	50	100		200 km
0	30	60		120 miles

OCEAN

Angel Falls

The highest waterfall in the world, nearly 1 km tall, named after the US pilot, Jimmie Angel, who discovered it. He crashed his plane on the top of the mountain and climbed down for 11 days to the Falls' bottom

SOUTH AMERICA

SOUTH AMERICA

Area
With an area of 17.8 million km², South America is slightly larger than Russia, encompassing around 12% of the Earth's landmass

Population
The population of South America is roughly 60% that of Europe – it's the 3rd least populated continent after Oceania and Antarctica

South America (5.68%)
422,535,000 people

$

Wealth
The nominal GDP of South American states is $3.9 trillion, slightly greater than Germany's ($3.5 trillion) and around 5% of the value of the world's economy

Countries
South America has 12 sovereign states, the least of all the inhabited continents. It also includes 3 non-sovereign territories: French Guiana (France), Falkland Islands (UK) and South Georgia and South Sandwich Islands (UK)

Highest capitals
3 of the 4 highest capitals are in South America: La Paz (1st, 3,640 m), Quito (2nd, 2,850 m) and Bogotá (4th, 2,625 m)

Coastal population, inland wilderness
The majority of the continent's population is concentrated along the west and east coasts, while most of the interior is taken up by the sparsely populated natural treasures of the Amazon rainforest in the north and Patagonia in the south

Inca Empire
Even without a system of writing, wheeled vehicles or the use of animals to power production, the Incas had an immense knowledge of advanced mathematics, astronomy and architecture

In the 16th century, before the European conquest, the Incas were probably the largest empire on Earth, stretching from what is now Colombia to central Chile, with a multi-million population

From rocks to life
The erosion of the Andes led to the sedimentation of rich elements which fertilized soils, crucial for development of pre-Columbian civilizations

Punta Pariñas ★
Westernmost point in mainland South America
81°19'43"W

Americas' oldest civilization
Norte Chico developed in what is now Peru, around 3500 BC. It lacked ceramics and visual arts, but it manifested itself through monumental architecture and earthwork designs

Nazca Lines
The most extensive and impressive geoglyphs on the planet and some of the greatest enigmas of archaeology. Probably created for ritual and astrological purposes between 500 BC and AD 500

"Motorcycle Diaries"
Ernesto "Che" Guevara's 1952 trip around the continent, during which he witnessed social injustices and capitalist exploitation. The experience of this trip had a big impact on his life as a revolutionary

Longest mountain range
At around 7,000 km long, the Andes are the longest terrestrial mountain range on the planet. 38 South American peaks are higher than Denali (6,190 m), the highest summit in North America

Punta Gallinas
Northernmost point in mainland South America
12°27'31"N

Highest single drop waterfall
Kaieteur Falls, Guyana

Highest waterfall ★
Angel Falls, Venezuela

Southern Hemisphere
Around 90% of South America's territory lies in the southern part of the globe

Equator

Amazon
South America's longest river and the world's largest by volume

Largest rainforest
The Amazon is the largest rainforest on the planet

Purus
Third longest river on the continent, 3,218 km long

Highest navigable lake
Lake Titicaca, Bolivia and Peru

★ Highest capital city
La Paz, Bolivia

Poverty
In South America, 11% of people live on less than $2 per day – a similar ratio to China

★ Driest non-polar place
Atacama Desert

Resources
Main mineral resources: petroleum, gold, silver, copper, iron ore and tin

Pampas lowlands
One of the world's main food-producing regions

Andean condor
The largest flying bird on the planet, by combined measurement of weight and wingspan (reaching 3.3 m)

Tropics
Most of South America lies in the tropics, north of the Tropic of Capricorn

PACIFIC OCEAN

Cabo Froward
Southernmost point in mainland South America
53°53'47"S

0	500	1,000	2,000 km
0	300	600	1,200 miles

Hummingbird

This family of small birds is native to the Americas, however the bee hummingbird, the smallest hummingbird and the world's smallest bird (5.5 cm long), lives in Cuba

Least visited
In 2015, Guyana was visited by only 207,000 tourists, less than any other country on the continent

Latin America
South American countries speak predominantly Romance languages. Only Guyana (English) and Suriname (Dutch) do not belong to this group

Alexander von Humboldt
Prussian geographer who, after travelling across South America, became the first man to describe the continent from a modern scientific point of view. Responsible for the first accurate drawings of Inca monuments and the first (1800) theory about human-induced climate change

Green anaconda
World's heaviest (97.5 kg) and 2nd longest (5.21 m) snake

Jaguar
3rd largest cat after the tiger and the lion, mainly found in South America

Brazil
Home to about half of the continent's population (207.8 million), nearly half its area (8,514,879 km²) and half of the value of its economy ($1.8 trillion)

Trans-Amazonian Highway
Around 5,000 km long, not fully paved. Both its construction and the logistical possibilities it opened, contributed greatly to deforestation in the Amazon

Ponta do Seixas
Easternmost point in mainland South America 34°47'35"W

Geographic centre
Cuiabá in Brazil lies at the centre of South America

Violence
South America is considered the most dangerous continent in the world, with 16 homicides per 100,000 people recorded in 2013

Richard Evans Schultes
He travelled the continent in search of a species of natural rubber that would end the US dependency on Asian rubber, unavailable due to the war with Japan during WWII. He was the first to alert the public to the Amazon's destruction

ATLANTIC OCEAN

Río de la Plata-Paraná
At 4,500 km in length, it is the 2nd longest river in South America

Río de la Plata-Paraná

SÃO PAULO
Population: 21.5 million

RIO DE JANEIRO
Population: 13.1 million

African slaves
It is estimated that between 1502 and 1866, up to 10 million African slaves were brought to South America and the Caribbean

Religion
Around 92% of South America's population is Christian

Tourism
Most popular destinations (visitors annually):
Brazil – 6.3 million
Argentina – 5.7 million
Chile – 4.5 million

Amazon rainforest
The Amazon contains over half of the planet's remaining rainforests. An estimated 390 billion trees, representing 16,000 different species, produce over 20% of the world's oxygen supply

BUENOS AIRES
Population: 15.5 million

Saharan fertilizer
The dust from the Sahara desert, which contains phosphorus, is an important fertilizer for the Amazon

Deforestation
Large scale deforestation started in the 1970s. Since then, around 20% of the Amazon has been destroyed. The process is attributed mainly to the clearing of land for cattle pasture

Southern Cone
Geographic region including Argentina, Chile and Uruguay (sometimes Paraguay and southern Brazil are included). It is characterized by the highest prosperity and standard of living in Latin America

Supporting life
One in 10 known species in the world lives in the Amazon rainforest. It also provides favourable conditions for the transmission of tropical diseases such as malaria and yellow fever

CARIBBEAN SEA

Lake Maracaibo
There is a persistent storm here with around one million lightning strikes per year

CARACAS

CUMANÁ
Oldest European settlement in South America (1521)

Our Lady of Peace monument
It's the world's tallest statue of the Virgin Mary and the tallest monument in South America, just centimetres taller than the Statue of Liberty (excluding the pedestal)

MÉRIDA
Heladería Coromoto is an ice cream parlour holding the Guiness World Record for offering 863 flavours

Oil fields
Venezuela's largest oil field is located entirely in the Orinoco Basin

World's largest oil reserves
Venezuela holds over 300 billion barrels of reserves in conventional oil as well as oil sand deposits equivalent in size to the world's entire conventional oil reserves

Beauty factory
Venezuela has won 22 titles in the Big Four international beauty pageants, the most successful country in this domain

Shortages and starvation
Because of state-controlled prices, Venezuelans have experienced shortages of basic goods such as food, water, medications and other staples

Orinoco

VENEZUELA

Little Venice
Its landscape reminded Amerigo Vespucci of Venice, thus he called it "Veneziola", or "Little Venice"

Economic and political crisis
Despite possession of great natural wealth, Venezuela has been a poorly governed state, heavily indebted, with exploding inflation rates and soaring poverty indices

Violence
Venezuela is one of the most dangerous places on Earth. US and Canadian diplomats are obliged to travel in armoured vehicles here

Angel Falls
The highest waterfall in the world, nearly 1 km tall, named after the US pilot Jimmie Angel, who discovered it. He crashed his plane on the top of the mountain and climbed down for 11 days to the Falls' bottom

Price of water
In 2013, due to a water shortage, water was more expensive than oil

Hyperinflation in Venezuela

500%
400%
300%
200%
100%

1980 1985 1990 1995 2000 2005 2010 2015

Orinoco river
4th largest river in the world by water discharge, after Ganges-Brahmaputra

Orinoco

Food crisis
Due to the food crisis, in 2016 alone, around 75% of the population lost on average 8 kg of body weight because of malnutrition, and many families were forced to give up their children for adoption to survive

Giant otter
Up to 1.7 m in length, it is the longest species of the weasel family. Once widespread in South America, today only about 5,000 giant otters survive in the wild, mostly in the Guianas

Venezuelans living in poverty

80%
60%
40%
20%

poverty
extreme poverty

1999 2005 2010 2015

Hugo Chávez
The youngest president in Venezuela's history. Described himself as a Marxist. He held a weekly radio broadcast in which he answered phone calls from the public

Criticized for Bolivarian propaganda and enabling acts, which allowed him to rule by decree in key public matters. He died in 2013, two months after he was supposed to have been sworn in as president for the 4th time

Death penalty
In 1863, Venezuela became the 1st independent state to ban capital punishment for all crimes, including serious military offences such as treason

0   75   150   300 km
0   50   100   200 miles

Angostura bitters

Concentrated liquor with herbs and spices used as an ingredient in cocktails. It comes from Venezuela but today is produced in Trinidad and Tobago. Its exact formula is a secret, split into 5 parts, each known to one person

Columbus – Third voyage

Columbus's third voyage (1498) during which he visited the continent of South America for the first time. It took 2 months to sail across the Atlantic

Columbus – Third voyage

Demerara sugar

A type of cane sugar originally cultivated in Demerara, an historical region which was a Dutch colony from 1745 to 1815. Today it is Guyana's smallest and most populous region

Smear campaign

In the late 19th century, white-sugar industry representatives started a successful smear campaign against brown sugar, showing repellent but harmless microbes that are in brown sugar

Orinoco Delta swamp forests

Beautiful, inaccessible ecoregion, covered mainly with permanently flooded rainforest, unsuitable for farming or logging

★ Jonestown

In 1978, 909 American sect members committed a mass suicide here

GEORGETOWN

Smallest

Although Suriname is the smallest independent state in South America, it is about twice the size of Austria

Space flights

French Guiana's location near the equator gives it an advantage for launching satellites due to the Earth's higher rotational speed at this point

Kaieteur Falls

The largest single-drop waterfall by volume of falling water, about 4 times taller than Niagara Falls

English

The only country in South America where English is the official language

Professor van Blommestein Meer

The construction of this reservoir flooded 1% of the country. It's one of the most inefficient hydroelectric projects by area flooded/MW generated

PARAMARIBO

European Space Agency

The agency launches its communication satellites from Guiana Space Centre

• CAYENNE

GUYANA

Claimed by Venezuela

SURINAME

The Dutch acquired Suriname in exchange for Manhattan Island in 1667

FRENCH GUIANA

Mount Roraima (2,810 m)

A tripoint, where the borders of three countries meet. There are 176 tripoints on Earth

★ Birds

Kanuku Mountains are home to the extremely rare harpy eagle – one of the largest raptors in the world – and 500 other bird species

Sranan Tongo

English-based creole language, which serves as a lingua franca for 80% of Suriname's population. The official language is Dutch

Overseas department

The largest department of France and the region with the highest GDP per capita in South America

Claimed by Suriname

Rainforests

In 2012, Guyana received a $45 million reward from Norway for its rainforest protection efforts

Claimed by Suriname

Low population density

Suriname and French Guiana have amongst the lowest population densities on the planet, just 3.2 and 2.1 people/km² respectively

Disputed territory

The area west of this disputed border is claimed by Venezuela. The dispute goes back to colonial times but the conflict returned in recent years when oil was discovered in Guyana's maritime areas on the Atlantic

Left-hand traffic

Suriname and Guyana are the only countries in continental South America with left-hand traffic

Claimed by Suriname

Suriname is in constant dispute about the borders with its neighbours. Only the border with Brazil is established

Suicides

Guyana has the 2nd highest suicide rate (30.6 per 100,000 people per year) in the world, after Sri Lanka

Pom

Famous festive dish in Suriname, originally cooked by Jews for Passover, its main ingredient is the root of the tropical plant taro

Quinoa
Peru is the leader in quinoa production. Between 2010 and 2014, its production rose by 280%

Peruvian wool
Considered the world's most luxurious fabric. It is very expensive because vicuñas which produce it, can only be shorn every three years

Chicama ★
World's longest left-hand wave that offers surfers a ride of up to 2.2 km long for over 2.5 minutes

Ticlio
Mountain pass with one of the highest asphalted roads in the world (4,818 m)

National University of San Marcos
Lima's university is the oldest university in the Americas (1551). Its professors and students played leading roles in Peru's history

P A C I F I C
O C E A N

The longest river?
The Amazon is in competition with the Nile for longest river (still debated)

Amazon river basin
The river has over 1,100 tributaries, 2 of which are longer than Europe's longest river and at least 18 of which are longer than 1,000 km

The river produces 20% of all the world's freshwater discharge to the ocean

Its discharge is greater than the next 7 largest rivers added together. In one year it could fill entirely the basins of Lake Ontario and Lake Michigan combined

The Amazon once drained to the Pacific Ocean instead, but it gradually reversed towards the Atlantic with the rise of the Andes

Cocaine production
Almost all cocaine is produced in Colombia, Peru and Bolivia

Rubber plants
Discovered in the 18th century. They were a game-changer for transportation and many other industries

IQUITOS
Centre of Ayahuasca tourism

Incan Empire
The largest empire in pre-Columbian America was mainly located in modern-day Peru

Amazon river
World's largest river by water flow. Its drainage basin covers 40% of the continent and accounts for around 20% of the world's total river flow

Isolated tribes
There are over 100 tribes in the Amazon that live without any contact with civilization. Due to their isolation, they lack immunity to common diseases, and any contact with outsiders can lead to the death of entire tribes

Ayahuasca
Famous rainforest vine with hallucinogenic properties. Used as a religious sacrament under the guidance of a shaman

Biodiversity
Brazil is the most diverse country on Earth. Around 25% of the world's rainforests grow here

PERU

Amazon

Potato
Spaniards brought potatoes from Peru in the 16th century

LIMA

36 languages
The Constitution of Bolivia recognizes 36 official languages besides Spanish

Lake Titicaca
The highest navigable lake in the world (3,812 m)

North Yungas Road
56 km long, named "world's most dangerous road". In 2006 up to 300 people died there

LA PAZ
Seat of the government. The highest capital city in the world (3,640 m)

BOLIVIA

Brazil nut
Bolivia is the main exporter of Brazil nuts, not Brazil

• SUCRE
Constitutional capital

Salar de Uyuni
World's largest salt flat. It has a similar size to Lebanon

Shortest women
Bolivian women have the world's lowest average height, 142.2 cm

Simón Bolívar
Bolivia's name comes from this military figure who led Bolivia, Colombia, Ecuador, Peru and Venezuela to independence. He was also the first president of Bolivia

0   200   400   800 km

0   125   250   500 miles

Pororoca

During intense rains the river Amazon creates a tidal bore, the Pororoca, with waves large enough to surf

Amazon

Amazon river

Enters the Atlantic Ocean in an estuary about 325 km wide

Only 7% of land is arable but the country is very productive agriculturally

Second largest soybean producer, after the US

Agriculture in Brazil

Initially, plantation owners tried to use locals for sugar production, but encountered resistance and diseases that affected the local population, forcing them to import African slaves

Brazil produces around a third of the world's coffee

Leader in the world's orange and orange juice production

Sweet water dolphins

Live deep in the jungle-covered basin of the Amazon. They are almost blind and use mostly their natural sonar to communicate and navigate

Jesus Christ lizard

It can run on water, never submerging its body more than a few centimetres deep. An average human would have to run more than 100 km/h to replicate this trick

Football

Brazil has won the FIFA World Cup 5 times, more than any other country

Brazilian pygmy gecko

The size of a fingertip. It's very light and its skin is water repellent so it can stand on water

"Little Japan"

Brazil has the greatest number of people of Japanese ancestry outside of Japan

"African connection"

The continents of South America and Africa were joined together until the supercontinent Pangaea broke apart about 225 million years ago. Therefore, they share similar fossils and rock layers

Crime

Considered one of the most dangerous countries in the world. The homicide rate is around 24 deaths per 100,000, about 6 times higher than in the US

Prisons

The prison system is heavily overcrowded (3rd largest prison population in the world). Prisoners can reduce their sentence by reading books (1 book = 4 days less)

BRAZIL

5th Largest country by area
Largest country by population

Bolsa Familia Program

If families ensure that their children attend school and get vaccinated, they receive financial aid

Pantanal

The largest tropical wetland in the world. When the dry season comes and the water level falls, it has the largest concentration of crocodiles on the planet, as they become trapped in isolated ponds

"Ordem e Progresso"

The country's motto, "Order and progress", is written on its flag

Low immigration

Only 2% of the Brazilian population is foreign-born

SALVADOR

Described as the largest African city outside Africa as 80% of the 3.7 million population can trace their ancestry to slaves

BRASÍLIA •

The country's capital took just 41 months to build (1956 to 1960)

Brazilian Gold Rush

Started in 1695. Brought approximately 400,000 Portuguese and 500,000 African slaves to work in the region of Minas Gerais

ATLANTIC OCEAN

Economic growth

By 2032, the economy of Brazil will overtake Germany's in size

The region of São Paulo contributes over 30% of the country's GDP (comparable to the GDP of Saudi Arabia)

The capital of the Kingdom of Portugal between 1815–22

RIO DE JANEIRO

Instituto Terra: reforestation

Since the 1990s, famous photographer Sebastião Salgado and his wife have planted over a million trees on nearly desertified land

SÃO PAULO ○

6th largest city in the world (21.5M people)

Carnival

Takes place 46 days before Easter. 80% of the annual consumption of beer and 70% of foreign visitors come during this time

Santos Basin

Location of Brazil's major oil fields

Iguaçu Falls

World's largest and probably most impressive waterfall system is shared by Argentina and Brazil

Paraná

Moconá Falls

3 km long, running parallel to the river. For 150 days per year, when the water level is high, the waterfall is not visible

Oil exploration

Between 2010 and 2014, 63% of the world's deepwater oil was discovered off the coast of Brazil. It was named "the 2nd independence for Brazil"

Praia do Cassino

World's longest beach (245 km)

Paraná

Paraguay

PACIFIC
OCEAN

"Silver country"
The name Argentina comes from the Latin word "argentum" meaning silver. Numerous expeditions set out to find the mythical "Silver Mountains". All of these failed but the reference to silver remained

Beef
Argentina is famous for its high quality beef. It's the world's 3rd biggest producer of this meat and has the highest consumption per capita, 65 kg/year

First plants
The ancestors of all plants which we have today appeared 472 million years ago. They were liverworts, very simple plants which lack stems or roots. Their fossils were discovered in Argentina in 2010

Dakar

Yerba mate
These leaves, rich in caffeine, are considered a "national infusion" in Argentina and neighbouring countries

Paraná

Banned "Messi" name
In Rosario, Argentine soccer star Lionel Messi's home town, naming newborn children "Messi" became so popular that the city's officials had to ban it

Río de la Plata

ROSARIO

Dakar

Atacama Desert
The driest non-polar desert in the world. Some parts of it haven't seen a drop of rain since records began

Escondida mine
World's largest copper mine by reserves

Nevado Ojos del Salado (6,908 m)
The highest active volcano in the world

6,688 m
Highest altitude reached with a car, a Suzuki Samurai, on the slopes of Nevado Ojos del Salado

2011 crisis
During the 2011 economic crisis, Argentina had 5 presidents in just 10 days

Dakar Rally
Route of the 2017 Dakar race 317 vehicles competed for 13 days on a 8,786-km-long route

Atucha plant
First nuclear power plant in Latin America, began operating in 1974

Avenida 9 de Julio
Buenos Aires hosts the world's widest avenue with 14 lanes

BUENOS AIRES
Tango was born here. It was influenced by both African and European

A N D E S

Cerro Aconcagua (6,961 m)
The highest mountain peak outside of Asia

Dakar Rally

CHILE

ARGENTINA

SANTIAGO

Width
Chile is only 350 km wide east to west at its widest point

NASA testing
Soil samples from the Atacama Desert are very similar to those taken from Mars. NASA used to test its materials and instruments for future missions here

Richards Deep
Deepest point in the Peru-Chile Trench and one of the deepest in the oceans (8,065 m)

Copper power
Chile provides 25% of the global copper supply

"Chilli"
The name Chile may come from the indigenous word chilli, which means "where the land ends"

The Church of the Company Fire
This fire in Santiago in 1863 resulted in the highest number of casualties in a building fire in history (up to 3,000 people died)

ROBINSON CRUSOE ISLAND (Chile)
Was home to the marooned sailor Alexander Selkirk (1704–9), and is thought to have inspired novelist Daniel Defoe's fictional character

PACIFIC
OCEAN

Friend's Day

Buenos Aires has a special holiday on 20 July called Friend's Day (El Día del Amigo), which was founded by Dr Febbraro, an Argentine dentist and philosopher, after he felt connected to everyone on Earth following the Apollo 11 moon landing

A T L A N T I C   O C E A N

Cosmetic surgery

Argentina has the highest rate of people who undergo plastic surgery (3%)

Psychoanalysis

Argentina has around 200 psychoanalysts per 100,000 people, the highest in the world

Hyperinflation

Since 1970, Argentina has had to "erase" 13 zeros off its currency as it has suffered from hyperinflation multiple times

Political beer

Some Argentinian political parties have their own brands of beer

Cocoa crisis

In 2000, a fungus spread throughout South America's cocoa plantations, devastating the economies of the region and driving up the price of chocolate

Andes

The world's longest continental mountain range stretching 7,000 km and the highest mountain range outside Asia. It extends through 7 countries

S
A
N
D
E
S

Inhumanity

In 1881, 11 South Patagonian locals were kidnapped to be displayed like animals in Paris and Berlin zoos

Titanosaur ★

In 2014, a farmer found a bone the size of an adult human. The discovery led to the excavation of the largest land animal that ever lived on Earth. 40 m long, 20 m high, and an estimated 77 tonnes in weight

People of water and stars

Indigenous people of Southern Patagonia ritually painted cosmic patterns on their bodies, believing that people originate in the stars and become such after death

For over 10,000 years they lived off the ocean, travelling from fjord to fjord in traditional canoes

FALKLAND ISLANDS (U.K.)

Patagonia

1 million km² of pristine and breathtaking lands including 7 national parks with crystal clear lakes, rivers, snow-capped volcanoes and mountains. Population density is just 1 person/km²

Perito Moreno

One of the last glaciers still growing. It's 60 m high, light blue in colour, with a rugged ice front moving slowly towards Lake Argentino

Strait of Magellan

The most important natural passage between the Atlantic and Pacific Oceans, first navigated by Ferdinand Magellan in 1520. Unstable weather conditions, narrowness and strong currents make it a challenging route

Falklands War

In April 1982, Argentine forces temporarily occupied the islands. British governance was restored two months later at the end of the war. Argentina still claims sovereignty of the Falklands

Ushuaia is the southernmost city in the world with around 57,000 citizens

USHUAIA

Cape Horn

These waters are particularly hazardous owing to strong winds, large waves, strong currents and icebergs. These dangers have made it notorious as a sailors' graveyard

Tierra del Fuego

Great Chilean Earthquake

World's greatest earthquake in recorded history (1960) with a 9.5 magnitude. 25% of all global seismic energy, in the 1906–2005 period, was released during this catastrophe

Diverse climate

Chile stretches over 10 major climate subtypes, according to the Köppen climate classification, from humid subtropical to alpine, tundra and glaciers

Divorce

Divorce was legalized as recently as 2004 in Chile and 1987 in Argentina. While in Chile the divorce rates are among the world's lowest, in Argentina they're the highest

Length

Chile is one of the longest north-south countries, stretching over 4,300 km

Southern Patagonian Ice Field

The largest freshwater ice reservoir on the planet outside Antarctica and Greenland. One of the last remnants of the recent ice age

Southern Patagonian Ice Field expeditions

The first European expedition to the Southern Patagonian Ice Field was in 1913

The first north-south crossing of these lands was accomplished in 1998

Many areas in Patagonia still remain unexplored and pristine today

Tierra del Fuego

"Land of Fire" named by Ferdinand Magellan who was the first European to see the fires lit by native tribes who inhabited the archipelago. At that time there were 8,000 indigenous people living in 300 "family canoes"

0   100   200
0   50   100   200 miles
0   100   200   400 km

48

Main areas of coca cultivation (2005–14)

Coca
A stimulant herb, used in the Andes since the ancient Inca Empire era. Consumed in a tea or by chewing, it is known for improving mood, restoring energy and is used to alleviate altitude sickness

Cocaine Cola
Until 1903, Coca-Cola included 9 mg of cocaine per serving

Cocaine market
The value of today's cocaine market has reached $500 billion annually, spent by over 20 million users

Colombia produces around 70% of this drug, while the US is the largest consumer of it

1 kg of the product sells for $50,000, while the cost of its production is only $1,500

Emeralds
Around 2/3 of these green precious gems, sold around the world, originate from Colombia

Coffee
Although coffee was unknown in South America until the late 18th century, today, Colombia is the world's largest exporter of arabica beans and 3rd largest producer of coffee

Biodiversity
The most biodiverse country per km². It's the 2nd most biodiverse country after Brazil, which is roughly 10 times greater in size

○ MEDELLÍN
It was the main hub of Pablo Escobar's drug cartel. It used to be the murder capital of the world with an average of 17 people killed every day, back in 1991

GALAPAGOS ISLANDS (Ecuador)

Marine iguana
The only sea-going lizard on the planet has evolved separately from land iguanas for 10 million years

The islands are famed for their vast number of endemic species

Charles Darwin
His research in the Galapagos led to the groundbreaking theory of The Origin of Species

● BOGOTÁ

El Dorado
The expeditions searching for the mythical golden city since the late 16th century have led to the mapping of much of northern South America and the Amazon Basin

Birds
Colombia has more species of birds than all of Europe and North America combined

Spectacled bear
The only species of bear living in South America, and the last surviving species of short-faced bear. These peaceful animals inhabit high altitude forests and feed mainly on plants

COLOMBIA
Named after Christopher Columbus

Blue anole ★
World's only lizard coloured entirely blue. It's found only on Gorgona Island

Liquid rainbow ★
Caño Cristales river is famous for the incredible colours of its riverbed as they change to yellow, green, blue, black and red

Korean War
The only Latin American country which took part as an ally of the USA during war in Korea in the 1950s

UNESCO
Quito and the Galapagos Islands were among the first 12 UNESCO World Heritage sites established in 1978

A N D E S

Colombian conflict
On-going internal war in Colombia since 1964. In this period, 17% of Colombians have become direct victims of the unrest, with over 220,000 killed

The parties involved in the conflict include the Colombian government, extreme left and right-wing paramilitary groups and criminal syndicates controlling the multibillion dollar cocaine trade

Cinchona
Ecuador's national tree and source of quinine, the first known medicine for malaria

○ QUITO

ECUADOR
Means "equator" in Spanish

△ **Chimborazo (6,310 m)**
The most distant place from the centre of the Earth thanks to the ellipsoid shape of the planet

"Banana republic"
Most of the world's bananas are grown domestically in India but Ecuador is the largest exporter, (around a quarter of global exports)

Wikileaks
Ecuador granted asylum to its founder, Julian Assange, known for publishing secret information. He went in to hiding in the Ecuador embassy in London in 2012

"New World" (The Americas)		"Old World" (Afro-Eurasia)	
peanuts	peppers	bananas	citrus fruits
potatoes	cacao	grapes	honeybees
tomatoes	beans	onions	cattle
pineapples	tobacco	coffee	sheep
pumpkins	vanilla	wheat	pigs
corn	turkeys	rice	horses

Columbian exchange
The voyages of Christopher Columbus initiated a great transfer of species, ideas and diseases between the New and the Old Worlds

0	100	200	400 km

0	50	100	200 miles

Guarani & Spanish
93% of Paraguayans are descendants of indigenous Guarani women and Spanish men making it the most homogeneous nation in South America

Stevia
The indigenous Guarani have used this sweet herb for millennia. Nowadays, Paraguay is the 2nd largest producer of the popular sugar substitute

Paraguay river
It's the most important feeder of the Pantanal, the largest tropical wetland on the planet

Economy v ecology
There are around 50 dams already built in the Paraguay basin, and another 80 in the pipeline. Use of hydroelectric power brings benefits to the economy but may destroy the Pantanal ecosystem forever

Lapacho
Used in ancient herbal medicine, firstly by the Incas for treating infection, fever and stomach problems

PARAGUAY

31 presidents
Paraguay had 31 presidents between 1904 and 1954. Most of them were forcibly removed from office

"The Social Contract"
After overthrowing the Spanish, Paraguay's first dictator wanted to create a utopian society based on Jean-Jacques Rousseau's "The Social Contract"

War of the Triple Alliance
The deadliest conflict in the history of Latin America, fought between Paraguay, and Brazil, Argentina and Uruguay from 1864 to 1870. Almost 70% of Paraguay's adult males were killed

Territorial loss
After the War of the Triple Alliance, Paraguay lost around 140,000 km² of its disputed territory (the size of Tajikistan)

Soybean
Although originally native to East Asia, Paraguay is the 6th largest producer of soybeans which account for nearly 40% of the country's exports

Río de la Plata-Paraná
South America's 2nd longest river, 8th in the world

City founded by Japanese immigrants in 1961, famous for soy cultivation

ASUNCIÓN
Officially Nuestra Señora Santa María de la Asunción

COLONIA YGUAZÚ

Itaipu Dam
In 2016, it became the world's largest hydroelectric power plant by annual energy production

José Gervasio Artigas
This Uruguayan national hero spent his last 30 years of life as an exile in Paraguay

	Paraguay	Uruguay
Population	6.6 million (105th)	3.4 million (135th)
Rural population	40.3%	4.7%
Population in poverty	26.9%	12.4%
Median age	27.8 years	34.7 years
GDP per capita	$4,000	$15,700
Adult obesity rate	15.1%	27.6%

Progressive leader
Among South American countries, Uruguay is ranked 1st in: peace, democracy, lack of corruption, quality of living, prosperity, security and press freedom

Paraguay & Uruguay
Both countries have similar names, but no historical link exists between them

The humblest president
Uruguay's 40th president, José Mujica, (in office 2010–15) maintained proudly an austere, rural lifestyle, donating 90% of his public salary to charities. He declined to use the presidential palace or its staff

After completing his term in office, he returned to farming and cultivates chrysanthemums for sale

Renewable energy
95% of electricity in Uruguay is provided by renewable energy from wind farms and hydroelectric facilities

Tropics
The only country in Latin America which is entirely outside the tropics

Independence
After a struggle between Spain, Portugal, Argentina and Brazil, Uruguay gained independence in 1828

Uruguayan Spanish
Modified due to a large number of Italian immigrants who spoke Cocoliche, an Italian/Spanish pidgin

Josephoartigasia
Extinct genus of giant rodent, first found in Uruguay. It resembled a cow-sized guinea pig. It was named in honour of the national hero, José Artigas

Río de la Plata
River estuary, 220 km wide, formed by the conjunction of the Paraná and Uruguay rivers. Its coastal areas are the most densely populated parts of Uruguay and Argentina

Liberalism
Considered one of the most liberal countries. Marijuana, same-sex marriage and abortion are legal here

URUGUAY
Officially the Oriental Republic of Uruguay

Largest rodent
The largest rodent living today is the capybara, weighing 50 kg and measuring around 1 m in length. It is widespread throughout South American forests

First FIFA World Cup
In 1930, the inaugural FIFA World Cup was held in Montevideo. Uruguay was the 1st nation to win it by defeating Argentina 4–2

Montevideo's port is the most advanced container terminal in South America. Its rivalry with the port of Buenos Aires dates back to Spanish colonial times

MONTEVIDEO

Uruguay's livestock
The economy is strongly based on agriculture – cattle and sheep outnumber the human population by 4 to 1 and 3 to 1 respectively

Paraguay

Paraná

Uruguay

Paraná

Río de la Plata

75   150   300 km
50   100   200 miles

First civilization
Minoan civilization of Crete
is the oldest recorded European
civilization (3650–1400 BC)

EUROPE

EUROPE

Area

Europe has a comparable size to Canada, 9.91 million km² – 6.7% of the Earth's land area

Population

Europe is the only continent where the population is expected to start declining by 2022, and by 2100 it is estimated it will be smaller by around 100 million people

Europe (9.94%)
738,849,000 people

$

Wealth

The total GDP of European countries is $19.1 trillion – around 1/4 of the value of the world's economy

Countries

There are 44 countries in Europe plus the eastern parts of Russia and Turkey

ARCTIC
OCEAN

Norwegian Current

Vatnajökull

The largest glacier in Europe by volume and the 2nd largest by area. It's up to 1 km thick

ATLANTIC
OCEAN

Scandin

Warm currents

The climate of Europe is strongly affected by the North Atlantic Current which brings warm sea temperatures all the way from the Gulf of Mexico

North Atlantic Current

Industrial revolution

Started around 1760 in Great Britain. Innovations in steam power, iron making and textile production together with inventions in machinery led to machines being substituted for manpower

Population:
10.5 million
LONDON

Iron curtain

"Iron curtain"

Post WWII border defences and political and ideological barrier between the countries influenced by the Soviet Union to the east and NATO members to the west

Spanish Armada

In 1588, Philip II of Spain sent 130 ships to England planning to overthrow Queen Elizabeth I. Half of the fleet was destroyed in storms or fighting the English navy. It marked the beginning of Spain's fall as a global power

Spanish Armada

Migration period

There was widespread migration of Europeans between the 1st and 7th centuries AD, linked to the fall of the Roman Empire. The exact reasons for this remain a point of debate

Alps

Atlantis

The triple junction of tectonic plates used to be the location often associated with the mythical land of Atlantis

Catalan Atlas

The most important and arguably the most beautifully illustrated world maps of the medieval period, produced in 1375 by the Majorcan cartographic school. It shows some of the islands of the Azores which were not "officially" discovered for another 50 years

Roman Empire

At its height in the 2nd century AD, the empire covered half of Europe and comprised 21% of the world's population

Cabo da Roca

Westernmost point in mainland Europe
9°29'56" W

Canary Current

Relicts of evergreen Europe

Around 15–40 million years ago, Europe was covered with evergreen, laurel forests. Their last remnants can be found in the green v-shaped valleys of Madeira

Punta de Tarifa

Southernmost point in mainland Europe
36°00'15" N

MEDITERRANEAN SEA

World's third largest sea

Messinian Salinity Crisis

5 million years ago, the Mediterranean Sea dried up almost entirely, due to the closure of the Strait of Gibraltar. If the strait closes again, the sea will dry up in around 1,000 years

Cape Nordkinn
Northernmost point in mainland Europe 71°08'02" N

Ural Mountains
Easternmost point in mainland Europe 66°37'05" E

Largest empire in history
The British Empire at its height in 1920, held 24% of territorial land on the planet. In 1913, it included 412 million people, 23% of the world's population at that time

Territories that were once part of the British Empire

Rich countries
4 Nordic countries (Norway, Denmark, Sweden, Finland) are among the top 10 European countries with the highest GDP per capita

Eurasia division
The border between Europe and Asia was first set as the Ural Mountains in the 18th century

Ural Mountains

Volga

• MOSCOW
Population: 12.3 million

"General Winter"
Russia's winter climate has contributed to several failed invasions of this country: Swedish invasion of 1708, Napoleon's Russian Campaign, Hitler's Operation Barbarossa

Last glacial maximum
Around 24,500 BC the ice sheet was at its greatest southern extent. It melted 12 millennia later. At that time, the sea level was 125 m lower than today

Last glacial maximum

Dnieper

European Russia
Occupies almost 40% of the total area of Europe. 110 million people live here – 77% of Russia's and 15% of Europe's population

European Union
Political and economic union of 28 member countries, although the UK voted to leave in 2016. Its single market economy is based on free movement of people, goods, services and capital. It's the 2nd largest economy on the planet with a population of about 510 million

Carpathian Mts
Over one third of all plant species in Europe grow in this region

Carpathian Mts

Caucasus

Euro
Currency of the eurozone, introduced in 2002 and currently used in 19 European countries. In 2017 there were €1,109,000,000,000 in circulation

Turkish Straits
The strategic importance of the Dardanelles and Bosporus have always led to disagreement and wars in this area

Danube

Black Death
From 1346 to 1353 this pandemic killed between 30% and 60% of Europe's population

Balkans

"Ode to Joy"
Anthem of the European Union. It is the last movement of Beethoven's 9th Symphony, his final complete symphony, first performed in 1824

ISTANBUL
Population: 14.6 million

"Europa"
According to Greek mythology, Europa was a Phoenician princess and the mother of King Minos of Crete

Ancient Greece
Cradle of Western civilization

Cyprus
Despite the fact that this island lies closer to Asia than Europe, it is sometimes considered part of Europe

0	250	500	1,000 km

0	125	250	500 miles

Dutch Golden Age
During the 17th century, the Netherlands was a European leader in trade, art, science and military power

Holding back the sea
Area of the Netherlands below sea level

Dutch East India Company
This trade organization, established in 1602, was one of the first international corporations in the world. It was also the first company to issue shares

Caribbean territories
The Netherlands has 3 self-governing territories (Aruba, Curaçao and Sint Maarten) and 3 special municipalities (Bonaire, Saba and St Eustatius) in the Caribbean

Around 1/3 of the population lives in this area

Port of Rotterdam
The largest port (by cargo tonnage) in Europe and 4th in the World

Oosterscheldekering
The largest of the 13 dams built to protect the country from flooding

"Battlefield of Europe"
Territory of Belgium has served as the major battlefield during many wars including both World Wars

Begium surrendered to Germany during WWII after only 18 days. France surrendered one month later

Monarchies
All three Benelux countries are European monarchies, others being: Andorra, Denmark, Liechtenstein, Monaco, Norway, Spain, Sweden, UK and the Vatican

WADDENZEE

Holding back the sea
25% of land is below sea level, 21% of the population lives there

The process of reclaiming land areas from the sea started in the 16th century. It represents 17% of today's area

The name "Netherlands" means "low countries"

Canals
Over 4,000 km of navigable canals, rivers and lakes

Bicycles
18 million bicycles for 17 million people

Flowers
75% of the world's flower bulb production

Windmills
Of approx. 10,000 windmills that operated in the mid-19th century, 10% are still working

AMSTERDAM
There are more than 1,200 bridges over the Amsterdam canals

NETHERLANDS

THE HAGUE
Seat of the International Court of Justice

ROTTERDAM

Gay rights
1st country to legalize same-sex marriage (2001)

Country of compromise
Since the 19th century the country has been ruled by coalition governments

The country has a similar population size (16.9 million) to the Moscow metropolitan area

Dutch anthem
Het Wilhelmus is the oldest national anthem in the world. It was written in the 16th century but became an anthem in 1932

ANTWERP
World's diamond capital (80% of all rough diamonds and 50% of all cut diamonds pass through here)

The capital of Belgium hosts the headquarters of European Commission, NATO, World Customs Organization, Eurocontrol, Benelux General secretariat and many other organizations

BRUSSELS

Dutch-speaking

French-speaking

Languages
This line divides the population between the Dutch-speaking (59%) in the north and French-speaking (41%) in the south

BELGIUM

The Big Bang theory
Priest and physicist Georges Lemaître came up with this theory in 1927

Benelux
Politico-economic union of Belgium, the Netherlands and Luxembourg signed in 1944. 3 out of 6 founding members of the EU (others being Germany, France and Italy)

Independence
Kingdom of Belgium seceded from the Netherlands after the Belgian Revolution of 1830

Billiard balls
World's leading exporter of billiard balls (almost 80%)

Ardennes Forest
The location of the last major German offensive of WWII. A surprise attack resulted in America's highest casualties during the war

Luxembourg is a similar size to the smallest US state Rhode Island

GDP
Second highest GDP/capita after Qatar

Workforce
Almost 50% of the country's labour force commutes to work from outside the country

LUXEMBOURG

LUXEMBOURG

Official languages
French, German and Luxembourgish

0   25   50   100 km
0 10 20   40 miles

Channel Tunnel
50 km long, 75 m deep. Around 21 million passengers transported annually

In May 1940, 338,226 Belgian, British and French troops were surrounded by German forces and had to be evacuated to the UK by more than 800 ships

Dunkirk evacuation

DUNKIRK

Battle of France
During the Second World War France was expecting the German invasion of its heavily fortified Maginot Line. Instead, the Germans attacked through Belgium

The largest seaborne invasion in history:
6 June 1944. Postponed by a month | 160,000 Allied troops against 50,000 Germans | Liberation of Paris: 25 August 1944

1918 flu pandemic
Started in a hospital in Étaples and killed 75 million people

English Channel
No army has successfully invaded England across the English Channel since 1688

Normandy Landings

ROUEN

Omaha
Most heavily defended beach during D-Day

Paris–Rouen
First competitive motor race in 1894, 126 km in 6 h 48 min

Champagne
The term "Champagne" is reserved exclusively for wines that come from this region

Battle of Verdun
One of the largest and longest battles of WWI on the Western Front

Polo shirt
Designed by René Lacoste, 7-time Grand Slam tennis champion

PARIS
Population of 11.1 million people

Over 300 varieties of cheese are produced in France

Asterix the Gaul
The action of this famous comic book took place in present-day Brittany

24 Hours of Le Mans
World's oldest active sports car race held annually since 1923

Siege of Orléans
The watershed of the Hundred Years War between France and England. French army's first major victory, led by Joan of Arc

Maginot Line

The Loire Valley
The valley is home to many beautiful Renaissance châteaux

Louis XIX
He was a king for 20 minutes before he abdicated

Loire

STRASBOURG
Seat of the European Parliament and the Council of Europe

Bay of Biscay
Home to some of the Atlantic Ocean's fiercest weather conditions

Cognac
Variety of brandy produced exclusively in the region of Cognac

German Occupation

Vichy France

FRANCE

Cognac

Château Lafite Rothschild
Winery producing one of the most expensive red wines in the world

Vichy France
Territory of the French State established by collaborationist government after the German invasion in 1940

The Lumière brothers invented the cinematographe here

LYON

Mont Blanc
4,808 m. One of the "Seven Summits". First ascent in 1786

There is an 11-km-long highway tunnel under the mountain

Overseas territory
Around 20% of French territory lies outside Europe

No food waste
In 2016, France became the first country to prohibit supermarkets throwing away food

The borders of modern France are roughly the same as those of ancient Gaul

Second largest producer of nuclear electricity in the world after the US

BORDEAUX

La forêt des Landes
The largest maritime pine forest in Europe. Established in 18th century and managed entirely for industrial purposes

Millau Viaduct
World's tallest bridge (343 m)

Between 1309 and 1377, 7 popes resided here

MONACO

Tourism
France has the highest number of tourists in the world, 83 million annually

France is the largest country in the European Union, and occupies 14% of its total area

AVIGNON

Cannes Film Festival

Alps

Pyrenees

MARSEILLE
France's oldest city (600 BC)

Côte d'Azur

ETA
Terrorist, separatist group fighting for independence for the Greater Basque Country. Responsible for over 800 deaths since 1968

The Declaration of the Rights of Man and of the Citizen, world's first declaration of human rights, was passed in France in 1789

Immigration

There are over 3 million French people of Algerian descent

0 — 50 — 100 — 200 km
0 — 25 — 50 — 100 miles

CORSICA
The most mountainous island in the Mediterranean, with 120 peaks over 2,000 m

AJACCIO
Birthplace of Napoleon Bonaparte (1769)

Hannibal
In 218 BC Hannibal crossed the Alps with an army of over 50,000 men and 40 elephants

South Tyrol
Autonomous province where 69% of the population speaks German as their first language

Pasta lovers
Italians consume ç. 25 kg of pasta annually per person

Mont Blanc
(4,808 m)

MILAN

First newspaper
Created in Venice in 1566, it was handwritten

Verona Arena
Amphitheatre famous for large-scale, spectacular opera performances in the ancient surroundings

VENICE
The Venice Film Festival is the oldest in the world

Museums
There are over 3,000 museums in Italy

Shroud of Turin
Believed to be the burial shroud of Jesus. Its origin is unknown and authenticity debated

PARMA
Home of Parmesan cheese. Aging time from 1 to 3 years

UNESCO
Italy has the highest number of UNESCO World Heritage sites – 53

GENOA
1st modern bank (12th century)

Imola
Famous F1 driver Ayrton Senna died in an accident on this track in 1994

The oldest state
San Marino claims to be the world's oldest surviving sovereign state (established AD 301)

SAN MARINO

Establishment
This small country was established by Marinus, a Christian stonemason who sought freedom from religious persecution by the Roman Empire

PISA

FLORENCE
The Renaissance, the cultural revival after the Middle Ages, began here in the 14th century

LIGURIAN SEA

SIENA
Today Siena has around 50,000 inhabitants, roughly the same as before the Black Death in 1348

North
South

Traditional division between the rich North and poor South regions of Italy

Costa Concordia
Location of the sinking of this famous cruise ship. It took 2 years to remove it

ITALY

ADRIATIC SEA

VATICAN CITY

ROME

Fascism
This nationalist movement was born in Italy during WWI

Via Appia
The most important road of Ancient Rome connecting the capital with the naval port in Brindisi

ASINARA
Home to a colony of 250 wild albino donkeys

190 km – the shortest distance between Sardinia and the Apennine peninsula

Via Appia

NAPLES

BRINDISI

Pizza was invented in Naples at the end of the 19th century

Mount Vesuvius
Its famous eruption in AD 79 destroyed the historic city of Pompeii

SARDINIA
Often called the "open museum" due o high number of archaeological sites

Sardines were named after Sardinia

Amalfi coast
Full of small, beautiful towns squeezed between mountains and the sea

GULF OF TARANTO

TYRRHENIAN SEA
Named after the prince Tyrrhenus who founded the Etruscan League of twelve ancient Italian cities around 600 BC

Buen Ayre
The capital of Argentina, Buenos Aires, takes its name from this hill in Cagliari

Strait of Messina
Location of the Scylla and Charybdis monsters from Greek mythology

USTICA
Fascist prison island

Messina earthquake
The epicentre of the 1908 earthquake which killed up to 200,000 people. The shock lasted 30–40 seconds

Archimedes was born in Sicily

PALERMO

Christopher Columbus, Marco Polo and Amerigo Vespucci were all born in what is now Italy

Etna
Europe's tallest (3,323 m) active volcano and one of the most active volcanoes in the world

SICILY
The biggest island in the Mediterranean

The island is full of Greek ruins. It was acquired by the Roman Republic in 241 BC

Languages
Although historically connected with Italy, the official languages of Malta are English and Maltese

50   100   200 km
25   50   100 miles

Allied invasion of Sicily
Codenamed Operation Husky. Started on the night of 9 July 1943. For the first time during WWII the Allies relieved the eastern front and forced Hitler to shift significant powers to Italy

Fishing boats
Traditional Maltese fishing boats have a pair of eyes painted on the bow, representing the eyes of Egyptian god Osiris. This custom goes back to ancient Phoenician times and is supposed to protect the fishermen at sea

MALTA

VALLETTA

0 100 200 km
0 50 100 miles

AZORES (Port.)

These 9 green, volcanic islands lie on 3 tectonic plates

N. American Plate
African Plate
Eurasian Plate

PONTA DELGADA

MADEIRA (Port.)

Cristiano Ronaldo
This famous football player was born in Funchal in 1985

Cabo Girão FUNCHAL
Cliff with the highest skywalk in Europe (580 m)

CANARY ISLANDS (Sp.)

Albino crabs
These blind, white creatures can only be found on Lanzarote island

LANZAROTE
LA PALMA TENERIFE
LA GOMERA LAS PALMAS
EL HIERRO GRAN CANARIA
FUERTEVENTURA

Teide volcano
The highest point above sea level in the Atlantic Ocean (3,718 m)

Language
Portuguese is the sole official language of 9 countries: Portugal, Angola, Brazil, Cape Verde, East Timor, Equatorial Guinea, Guinea-Bissau, Mozambique, and São Tomé and Príncipe

Over 250 million people speak Portuguese

First global empire
In the 15th century, Portugal became the first colonial empire, controlling territories that today belong to 60 countries

Slavery
In 1761, Portugal abolished slavery

ATLANTIC OCEAN

Vasco da Gama

Vasco da Gama
In 1497, Vasco da Gama left Lisbon with a fleet of 4 ships to reach India by sea for the first time in history

W N E S

Columbus

Magellan

The UK and Portugal have been allies longer than any other countries in the world (since the 14th century)

Fado
This traditional Portuguese music genre is classified as world cultural heritage by UNESCO

NAZARÉ
World's biggest wave surfed (30 m high)

LISBON

"Our Lady of Fátima"
Blessed Virgin Mary allegedly appeared here to three children 6 times in 1917

Vasco da Gama Bridge
Longest bridge in Europe (12.3 km)

Earthquake & tsunami
In 1755, Lisbon was destroyed by an earthquake and tsunami killing up to 90,000 people

Portugal is the world's 5th most peaceful country

PORTUGAL

OPORTO
The city's ancient name "Portus Cale" is the origin of the country's name

Border
The border between Spain and Portugal is referred to as "The stripe". It has hardly changed since the signing of the Treaty of Alcañices in 1297

Renewable energy
In May 2016, Portugal was powered for over 4 days exclusively by renewable energy

Muslims occupied Spain for 800 years. This line represents the greatest northern extent of Muslim rule

Muslim occupation

SANTIAGO DE COMPOSTELA
Although it is capital of the autonomous community of Galicia, it is not capital of the province it is situated in (A Coruña), which is a smaller administrative unit

Camino de Santiago
The main route of the pilgrimage to the reliquary of the apostle St James the Great. In 2014, 200,000 people participated in this pilgrimage

Don Quixote
Written by Cervantes in two parts (1605 and 1615). It is considered to be the first Western nov...

SALAMANCA
The oldest university in Spain (1134) and the world's third oldest in continuous operation

Football
Spain has dominated world football in recent years, winning two European Championships (2008, 2012) and one World Cup (2010) in the 21st century

Cork
Spain and Portugal provide around 80% of the world's cork

Spanish Inquisition
Institution converting and punishing heretics. Between 1478 and 1834 approximately 150,000 people were detained and over 3,000 executed

OLIVENÇA/OLIVENZA
The two countries have argued over the ownership of this city since 1815

Cannabis cultivation and possession is legal here

Heat of Andalusia
Seville is the warmest city in continental Europe with an average temperature of 19.2 °C, but the hottest summers are in Córdoba, where temperatures frequently exceed 40 °C

CÓRDOBA

SEVILLE

PALOS DE LA FRONTERA

"Air-ground intersection"
There is a road in Gibraltar which crosses the airport runway. When a plane lands, cars must stop for the red light

Gibraltar
British overseas territory with its own autonomous government. The currency is the pound sterling. In the 2016 "Brexit" referendum, 96% of Gibraltarians voted to remain in the EU

Costa de la Luz

Costa del So...

Entrance to the Mediterranean is 13 km wide

CEUTA
Spanish autonomous city

El Castillo
One of the oldest cave paintings in the world (40,000 years old)

Camino de Santiago

Guernica ★
Bombing of Guernica in 1937 was an inspiration for Picasso's famous painting

PAMPLONA
San Fermin festival includes the famous Running of the Bulls

€ Although Andorra is not in the EU, the euro is its official currency

⚔ Andorra hasn't been at war for almost 1,000 years

ANDORRA

ANDORRA LA VELLA

Cocaine
In 2014, Spain had the 2nd highest prevalence of cocaine use (2.3% of the population) after Scotland

Civil War
Spain stayed neutral during both world wars, but the Spanish Civil War (1936–39) killed more than 500,000 people

P y r e n e e s

Sagrada Familia
The construction of this monumental church, designed by Gaudí, began in 1882, and is planned to be completed by 2026

Costa Brava

Catalonia
Autonomous community of Spain with its own language and strong political and cultural autonomy

❄ During the last ice age, the ice didn't reach Spain

World War II
$ British MI6 spent up to $200 million in bribes in order to keep Spain from joining the Axis during WWII

BARCELONA

SPAIN
2nd largest country in the EU (after France) with a similar size to Thailand

Totalitarianism
Between 1939 and 1975, Spain was under the totalitarian rule of General Franco

Costa Dorada

BALEARIC SEA

MADRID

Sobrino de Botín, founded in 1725, is one of the oldest restaurants in the world. Famous painter Francisco de Goya used to work here as a waiter

BALEARIC ISLANDS

MINORCA
Name means "smaller island" (compared to Majorca)

Holy Grail
The Golden Chalice, from which Jesus was supposed to have drunk at the last supper, is kept in Valencia Cathedral

Costa del Azahar

MAJORCA
☠ Majorca and Minorca used to be important pirate bases during Roman times

La Mancha
World's largest wine growing region with over 190,000 ha covered by wineries

La Tomatina
Annual festival held in Buñol which involves a giant tomato fight. In 2015, almost 145,000 kg of tomatoes were used ★

VALENCIA

IBIZA
Prime spot for the nightclub party scene of Europe. Most of the world's best DJs have played here ♪

Over 400 million people speak Spanish which makes it the second most spoken first language after Chinese

Languages
There are 4 main languages spoken in Spain: Castilian, Catalan, Basque and Galician

Only 22% of the population speak English

End of the world
According to Nostradamus and his oracles, Ibiza will be the last refuge on Earth when the world ends

Olive oil
Spain is the largest olive oil producer with around 40% of the world's production. However the biggest exporter of oil is Italy

Costa Blanca

Decentralization
Spain is one of the most decentralized countries in the world

Spanish Costas
☀ The Mediterranean coast of Spain is famous for its diversity and beautiful beaches, and can be divided into 9 main sections

Transcontinental country
The only European state that borders an African country (two autonomous Spanish cities border Morocco)

Costa Calida

M E D I T E R R A N E A N   S E A

Costa de Almeria

osta Tropical

ALBORAN SEA
🐟 These waters are a transition area between the Atlantic and the Mediterranean and therefore contain a combination of their species

0   50   100   200 km

0   25   50   100 miles

Arctic Circle
North of this line, the sun does not rise for 24 continuous hours at least once during winter, and similarly does not set for 24 hours at least once during summer. The area within the Arctic Circle covers about 4% of the Earth's total surface and is home to c. 4 million people living in 8 countries

Saunas
5 million people and 2 million saunas – 99% of Finns take a sauna at least once a week

Forests
Finland is the most forested country in Europe – over 74% of its land area (c. 230,000 km) is covered by trees

FINLAND

Knivskjellodden
The northernmost point in continental Europe, visited each year by over 200,000 tourists during the short arctic summer

Inarijärvi

Kakslauttanen
Resort where you can rent a glass igloo and watch the northern lights through the roof

Independence
Finland gained independence in 1918. Before that, it was part of Russia and Sweden

ROVANIEMI
Santa Claus lives here. The Santa Claus Post Office (address: Santa Claus Village, FIN-96930 Arctic Circle), receives some 700,000 letters every year

Land of a thousand lakes
There are over 188,000 glacial lakes in Finland

"Citizen's wage"
The first country to introduce a universal income paid by the government. The experiment started in 2017

Archipelago
Finland is the second largest archipelago by number of islands

Gas prices in Norway are the highest in the world

Baltic ringed seal
Sea ice is the ringed seal's natural habitat for breeding, feeding pups and hunting

During the winter, in the northern part, the sea ice is up to 70 cm thick

Gulf of Bothnia

K-278 Komsomolets
Soviet submarine powered by a nuclear reactor carrying two armed nuclear warheads sank here in 1989. 42 crewmembers died. It still remains about 1.5 km below sea level

Moose
It's the most dangerous animal in Sweden, causing over 6,000 car accidents annually

About 100,000 moose are killed during the annual hunting season, to keep their average population size at 300,000

Arctic Circle

Seat belts
Volvo invented the 3-point seat belt in 1959 and made it available to other car makers for free. It saves 1 life every 6 minutes

Taxes
The Swedish word for tax is "skatt" which also means treasure

Swedes pay some of the highest taxes in the world and most are very proud of it

SAD
Seasonal Affective Disorder, or winter depression, affects c. 20% of Swedes

SWEDEN

Although legalised in 1944, homosexuality was still classified by law as an illness until 1979. In protest, Swedes were calling in sick to work, saying that they "felt gay"

Norway is a leader in freedom of press, democracy and corruption rankings

NORWEGIAN SEA
Its average depth is over 2 km because unlike most seas, its bed, rich in oil and natural gas, is not situated over a continental shelf

Climate change
Norway is ranked as the country least likely to be affected by climate change

Electricity
99% of the country's electricity comes from hydropower plants

In 2008, Norway donated US$1 billion to promote Amazon rainforest protection

Winter Olympics
Norway has won 303 Winter Olympic medals, more than any other country

TRONDHEIM
All the Norwegian kings have been crowned here since Harald Fairhair, over 1,100 year ago

Norwegian Current
Most of its flow comes from the shallower and much warmer Baltic Sea. Because of this, the Norwegian Sea, unlike other Arctic seas, never freezes

Norwegian Current

"The Troll Wall"
The tallest vertical rock face in Europe (1.1km)

FAROE ISLANDS
Autonomous country within the Kingdom of Denmark

TÓRSHAVN

There are only three sets of traffic lights on the islands for a population of 49,000

Saimaa — The largest lake in Finland and 5th largest in Europe. Its coastline extends over 13,700 km

Nord Stream Pipeline

Baltic ice extent, winter of 2008 — The warmest winter of the last 100 years

During an average winter, over 40% of the Baltic Sea surface is covered with ice

Nokia — From the 1990s until its decline in 2007, Nokia generated 25% of Finland's GDP and provided over 20% of all the country's corporation tax

The first country to make Internet access a legal right (2010)

HELSINKI

Vasa warship disaster — Built in the 17th century, it was the most glorious warship in the world, however it sank on its maiden voyage just 1.5 km from port

Largest European peninsulas

Scandinavian Peninsula 750,000 km²

Iberian Peninsula 582,000 km²

Balkan Peninsula 470,000 km²

Apennine Peninsula 131,000 km²

"Swedish Deluge" — Series of Swedish invasions of the Polish-Lithuanian Commonwealth in the 17th century. About a third of the Commonwealth's population lost their lives. Invaders destroyed and robbed 188 cities, 81 castles and 136 churches

BALTIC SEA

GOTLAND

ÖLAND

Sea of ice — During the past 300 years the Baltic Sea has frozen 20 times over its entire area, most recently in 1987

Least saline sea — Due to limited access to the ocean, combined with over 200 river estuaries and low average depth (55 m), the Baltic's salinity is only around 1%

Nord Stream Pipeline — Stretching across 1,222 km, from Vyborg (Russia) to Greifswald (Germany), it's the longest underwater pipeline in the world

Ancylus Lake — The Baltic Sea was once a freshwater lake, disconnected from the ocean (until around 6000 BC)

Vettisfossen — The tallest waterfall in Europe, 6th in the world (860 m)

Education — Over 30% of Norwegians have completed education at the post-secondary level, more than any other nation in Europe

NORWAY

Environment — The first country to establish a Ministry for the Environment (1972)

Internet — Every prisoner in Norway has internet access

OSLO

Nobel Peace Prize is awarded here annually

No advertising — In Norway and Sweden, advertising to children under the age of 12 is banned

English — 89% of people in Sweden speak English

Ericsson Globe — Stockholm is home to the largest hemispherical building in the world hosting the largest scale model of the solar system

STOCKHOLM

Known as "Venice of the North", the city lies on over 12 islands connected by 42 bridges

Digital money — 95% of transactions in Sweden are cashless, the highest in the world

Vänern — The largest lake in the EU and the 3rd largest in Europe

Vättern

Astrid Lindgren — World-famous author who wrote over 100 children's books which sold 144 million copies. She invented her most recognised character, Pippi Longstocking, to cheer up her daughter, Karin, when she was ill

Most of Denmark's Jews survived the Holocaust due to a massive evacuation to neutral Sweden

COPENHAGEN — Has more bikes than cars

DENMARK

Lego — 19 billion lego bricks are produced annually

The Danish royal family is the oldest European monarchy

Battle of Jutland — The largest naval battle of WWI. It involved 250 battleships and 100,000 men. It was the last battle in history of such scale fought primarily by battleships

The Danish flag is the oldest, continuously used national flag in the world, first flown in 1219

0 — 50 — 100 — 200 km

0 — 25 — 50 — 100 miles

ARCTIC
OCEAN

Polar bear
Its population on this
archipelago (approximately
3,000) is larger than the
human population

SVALBARD (Norway)

Prohibition of burial
Due to low temperatures,
organic matter does not decay
here. Buried bodies exposed to
climate warming could
potentially cause an epidemic

Hans Island
The island is claimed by
both Denmark and Canada

*Largest national
park in the world*
The Northeast Greenland National Park
protects over 970,000 km² of land. It was
the 1st natural reserve established by
Denmark (1974). It hosts 40% of the
world's muskox, and is home to polar
bears, walruses, arctic foxes, seals &
beluga whales

Aurora Borealis
One of the few inhabited places on
Earth where Aurora Borealis can be
seen during the day (in February)

Erik the Red
1st European who sailed to
Greenland around the year
AD 1000. His son was the
first man to sail to North
America. He came up with
the name "Greenland" to
lure settlers seeking green
lands and a better life

GREENLAND SEA
The first person to describe the
complex currents of this sea was
Fridtjof Nansen. He also led the
1st expedition which crossed
Greenland's interior in 1888

Glaciers
Glaciers in this region are
the fastest in the world, moving
up to 20 m per day.
Greenland has been covered with
an ice sheet around 2 km thick
for over 400,000 years. The ice
sheet and glaciers cover c. 80%
of the island and if they all
melted at once, the ocean level
would rise by 6 m

*Treading
on thick ice*
Maximum ice
thickness in
Greenland is
over 3,200 m

The Medieval Warm Period
Increased temperatures around AD 1000
allowed the Vikings to create temporary
settlements in Greenland

Oil
Many geologists
believe that
Greenland has
some of the world's
largest remaining
oil resources

JAN MAYEN (Norway)
Gravel is the only exploitable
natural resource on this island

Giant depression
The central part of Greenland
forms a depression up to 150 m
below sea level which is being
carved by glaciers

GREENLAND
(Denmark)

Gunnbjørn Fjeld
Highest mountain in all of the Arctic
Circle area (3,700 m). Rarely
climbed due to its isolation (60 km
from unpopulated coastal area)

Dependency
The largest dependent territory in the world. Its
head of state is Queen Margrethe II of Denmark

*First openly gay
Prime Minister*
Jóhanna Sigurðardóttir, elected
in 2009, was the first female
Prime Minister of Iceland and the
first openly homosexual head of
government in the world

Iceland lies on a volcano.
Natural warmth underground
is used to deliver heat to 85%
of all buildings

No roads or railways
Travelling between towns and
villages is possible only by plane,
helicopter, boat or dogsled

World's largest island
If Australia is counted as a
continent rather than an island

NUUK
Capital of Greenland
with around 17,000
inhabitants

The only place in the world where
an oceanic ridge passes above
sea level and literally tears the
land apart 2.5 cm every year

First Norwegian
settlers in AD 874

ICELAND

World's 18th largest
island. Similar in
size to Cuba

Income
Sealing, whaling and fishing
are main sources of income
for Greenland's inhabitants

The world's northernmost
capital of a sovereign state
REYKJAVÍK

Eyjafjallajökull
Its 2010 eruption forced almost all
European countries to close their air
space. The airline industry lost up
to $200 million a day

The least populated NATO
member and the only one
with no standing army

Ice sheet
It is estimated that
Greenland's ice sheet is
between 400,000 and
800,000 years old

North American Plate

Eurasian Plate

★ **Youngest island**
Surtsey island was created by a volcanic
eruption in 1963 and today almost 100
bird species live there

| 0 | 100 | 200 | 400 km |

| 0 | 50 | 100 | 200 miles |

| 0 | 50 | 100 | 200 km |

| 0 | 25 | 50 | 100 miles |

Gulf of Finland

"Baltic states"
Estonia, Latvia and Lithuania

BALTIC SEA

Atheism
Estonia is one of the least religious countries. Only 14% of the population associates with a religion

TALLINN
"The Silicon Valley of Europe", one of the places vying for the nickname

Basketball is the most popular sport in Estonia

Chess grandmaster
Paul Keres, one of the best chess players ever, was so popular that his funeral in 1975 was attended by over 100,000 Estonians, around 10% of the population

ESTONIA

Lake Peipus

Lake Pskov

Pulli settlement
The oldest human settlement in Estonia. The first evidence of its existence was a dog's tooth from 11,000 years ago

Skype was developed by an Estonian team

Online voting
Estonia is the world's first country to incorporate the system of online political voting

Gulf of Riga

Baltic Chain
In 1989, 2 million people peacefully protested against their political situation within the Soviet Union by holding hands, creating a 675.5-km-long human chain

"Baltic Tiger"
The term used to describe any of the three Baltic countries during their economic boom between 2000–07

Latvia and Lithuania are almost the same size (Latvia is only 611 km² smaller)

Latvia and Lithuania are each other's main trading partner

RIGA

LATVIA

Latvia lacks any significant natural resources except wood

Colonies
Duchy of Courland and Semigallia (modern day Latvia) used to have colonies in Africa and South America

Eurasian Economic Union
It is a political and economic alliance among former states of the Soviet Union, created on 1 January 2015. Its current members are: Russia, Belarus, Armenia, Kazakhstan and Kyrgyzstan

Polish-Lithuanian Commonwealth
The 1569 union between these two countries made them one of the largest and strongest European kingdoms for over three centuries

Forests
Latvia's forests, which cover around 52% of the country, are called "green gold" due to their significance to the economy

Baltic Chain

Geographic centre of Europe
The geographical central point of the continent is located in the town of Purnuškės

Dictatorship
Under Lukashenko's harsh rule, the country is often described by western countries as a dictatorship

LITHUANIA

End of communism
First country to declare independence from the Soviet Union on 11 March 1990

Alexander Lukashenko has served as the country's first and only president (1994 – today)

Belarus has the 2nd lowest Democracy Index in Europe after Russia. It's also considered the continent's worst country in terms of freedom of the press

VILNIUS
Described by Napoleon as "the Jerusalem of the North"

Capital punishment
The only European country that carries out the death penalty

Ice hockey
Ice hockey is the most popular sport in Belarus

BELARUS

Peat
Used as a fuel or fertilizer, it is one of the country's most valuable mineral resources

"Social parasite tax"
Belarus passed a law forcing those who worked less than 183 days in a year to pay a $250 fine

MINSK
In 1974, Minsk was awarded a Soviet honorary title "Hero City" for its heroic fight during WWII

Białowieża Forest
Primeval forest with the largest population of European bison (only around 800 still alive). Almost half of the species living in the forest, live on the decaying trees on the forest floor

Unemployment
According to the official government statistics, the unemployment rate in Belarus is below 1%, the lowest in Europe

Chernobyl disaster
Areas of the highest contamination after the catastrophe at the Ukrainian nuclear plant in 1986

Censorship in Belarus
Insulting the president is punishable by up to 5 years in prison, and criticizing him by up to 2 years in prison

During the past decades, several journalists and activists have been killed, while many have been detained or tortured

The last independent TV channel in Belarus, Euronews, was withdrawn from broadcasting in 2012

Banking system
Belarus has 31 banks. 30 of them are owned by the government

War casualties
The Byelorussian Soviet Socialist Republic was the part of the Soviet republic that suffered most during WWII

Pornography
In Belarus, pornography is illegal

CHERNOBYL

0	50	100		200 km
0	25	50		100 miles

NORTH SEA

176 destinations
Germans have the greatest freedom to travel of any nation. They can visit 176 territories in the world without a visa

Some of the greatest classical music composers were German: Bach, Beethoven, Wagner and many more

Unification of Germany
Signed in 1871 at the Palace of Versailles in France, creating a federation of 26 states

HAMBURG
Officially named "Free and Hanseatic City of Hamburg" as it was a free, imperial city in the 18th century

World's 3rd largest exporter and importer

Renewable energy
30% of energy produced in Germany is renewable. It has the 3rd highest wind power capacity after the US and China

Following the Fukushima nuclear accident in Japan and a series of protests, Chancellor Angela Merkel announced the closure of all nuclear plants until 2020

Lufthansa Group
World's largest airline by number of employees (124,306 in 2016)

Division of Germany
The country was divided into West and East Germany in 1945. It resulted in the "Berlin Blockade" (1948-9) of the Allied controlled part of the city

Berlin Wall
It was 140 km long. During the communist era, 5,000 people crossed it successfully and 139 were killed during attempts

● **BERLIN**
There are 1,700 bridges in Berlin, more than in Venice

Division of Germany

Ruhr district
The region gained prominence thanks to steel and coal production during 18th-century industrialization. Several large cities such as Dortmund, Essen and Duisburg, form one of the largest urban agglomerations in Europe

Berlin Zoo
The largest zoological garden in Europe, home to over 20,000 animals, visited by more than 3 million people annually. It is the world's most comprehensive collection of species

Ruhr district

University of Göttingen ★
45 Nobel Prize laureates studied or taught here

Reformation ★
In 1517, Martin Luther nailed his Ninety-five Theses to the door of All Saints' Church in Wittenberg and began the Protestant Reformation

Autobahn
12,993 km long, it's one of the most dense and longest controlled highways in the world, known for the lack of a federally mandated speed limit for cars and motorbikes

★ **Cologne Cathedral**
At 157 m, it is the highest twin-spired church in the world. It is also the most popular landmark in Germany, visited by around 20,000 people daily

GERMANY

Population
With 80.7 million people, it is the most populous country in the EU

Ties of blood
When World War I broke out, the rulers of the European superpowers involved in the conflict, Kaiser Wilhelm II of Germany, King George V of the United Kingdom and Tsar Nicholas II of Russia were all first cousins

MAINZ ○
The Gutenberg Bible was the first book printed in Western Europe (1455)

Oktoberfest
6 million litres of beer are consumed during the festival every year

Hugo Boss
Designed the official uniforms of the Nazi Party

Bavaria
One of the richest regions in Europe. Only 20 economies in the world have larger GDPs than Bavaria – the level can be compared to the economies of Saudi Arabia or Sweden. BMW, Audi, Adidas and other large companies are based here

Munich Massacre
Terrorist attack during the Summer Olympics of 1972, organized by Palestinian group Black September. 11 Israeli team members and a German police officer lost their lives

○ **STUTTGART**
Mercedes and Porsche were founded here

Danube

Weihenstephan brewery
The oldest brewery in the world, continuously in operation for nearly a millennium, first opened in AD 1040

Black Forest
Large forested mountain range known for its health resorts and mineral water sources

Neuschwanstein Castle
Built in the 19th century by King Ludwig II of Bavaria, it was an inspiration for Disney's Sleeping Beauty Castle

○ **MUNICH**

BALTIC SEA

Start of World War II
WWII started here when a German battleship opened fire on a Polish military depot

GDAŃSK

Johannes Hevelius
Born in Gdańsk, he created the earliest comprehensive maps of the moon in 1647

End of communism
Free elections in 1989 initiated the collapse of communism in Europe

Polish diaspora
Over 20 million Poles and people of Polish descent live abroad

Nobel Prize record
Marie Curie is the only woman to win 2 Nobel Prizes & the only person to be awarded in 2 different branches of the sciences (physics & chemistry)

Patek Philippe & Co.
Antoni Patek was a Polish innovator who co-founded one of the world's most prestigious watch companies together with Adrien Philippe. They met after Patek was forced to emigrate because of his involvement in the uprising against Russia's partition of Poland in 1830

Pope John Paul II
He was the first non-Italian pope since the 16th century and the first pope who prayed at an Islamic mosque or at the Western Wall. In total, he travelled over 1.1 million km to 129 countries

TORUŃ
The home-city of Copernicus, Polish astronomer and the first man to propose that the Earth revolves around the Sun

His findings, rejected by the church and extremely controversial at the time, were published just before his death in 1543

Partition of Poland
After 1795, Poland disappeared from the map for the next 123 years. It was forcibly divided between Prussia, the Russian Empire and Austria

Prussian Partition
Austrian Partition
Russian Partition
Russian Partition

WARSAW

POLAND

Warsaw Uprising
When the Poles fought the Nazis during the Warsaw Uprising in 1944, Stalin ordered the Soviet troops to stand still and watch

After the Nazis razed 65% of the city to the ground, they retreated west and the Soviets entered, claiming the city's "liberation"

ŁÓDŹ
Before World War II, it was a major European hub of the textile industry. None of the major textile companies survived the 1990s economic transformation after communism

Amber Road
Historic trade route connecting Northern and Southern Europe (10th century and later)

Today, Poland is the largest exporter of amber in the world

Bełchatów Power Station
2nd largest fossil-fuel power station in the world, responsible for 20% of the country's electricity production

In 1939, Poland had the largest Jewish population in Europe (c. 3.5 million), and their largest community of over 230,000 lived in Łódź

Chopin Competition
One of the most prestigious piano competitions in the world, commemorating the works of the famous Polish composer. He died from tuberculosis at the age of 39 in France. His heart rests in Holy Cross Church in Warsaw

Fallen power
In the 16th century, Poland was a leading European power with its territories reaching the Black Sea

Auschwitz-Birkenau
Major death camp built by the Nazis on Polish soil during WWII. Over 1 million Jews were killed here

Wieliczka Salt Mine
Had been operational continuously until 2007 for over 700 years, the longest working salt mine in the world

Danube

Amber Road

Amber Road

Comparison of areas before and after World War II

Just before WWII
1 September 1939

Germany   Poland

Today

Germany   Poland

| 0 | 50 | 100 | 200 km |

| 0 | 25 | 50 | 100 miles |

Countries by area and population

Switzerland
41,293 km² — 8.3 million

Austria
83,855 km² — 8.5 million

Czechia (Czech Republic)
78,864 km² — 10.5 million

Hungary
93,030 km² — 9.9 million

Slovakia
49,035 km² — 5.4 million

Bertha von Suttner
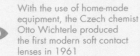
Austrian-Czech pacifist, the 2nd woman in history to receive a Nobel Prize and the 1st one to receive a Nobel Peace Price in 1905

She knew Alfred Nobel until his death in 1896 and it is believed that she convinced him to include in his will the award for promoting peace

Contact lenses
With the use of home-made equipment, the Czech chemist Otto Wichterle produced the first modern soft contact lenses in 1961

"Anschluss"
In 1938, Nazi Germany annexed Austria and parts of Czechoslovakia with the intention of creating a "Greater Germany"

Castles
There are over 2,000 castle in Czechia. Only Belgium and France have a higher density of such buildings

Swiss chocolate
The Swiss consume over 11 kg of chocolate per person per year, the highest amount in the world

Funeral business
Austrians have the highest spending on funerals per capita in Europe

Pilsner ★
Type of pale lager which originated in Plzeň in 1842. Today pilsner type beer accounts for over 2/3 of the world's be production

Best quality of life
According to the Quality-of-life index, Switzerland has the best quality of life

1st Nobel Peace Prize
Awarded in 1901 to the founder of the Red Cross, Henry Dunant

The EU Anthem
Ludwig van Beethoven, a German who lived in Vienna, was almost completely deaf when he wrote the music for "Ode to Joy", the official Anthem of the European Union

Inventor of LSD

Swiss scientist Albert Hofmann became the first person to synthesize and ingest the psychedelic substance known as LSD while trying to create a migraine medicine in 1938. He died at age 102

Women's right to vote

Liechtensteinerinnen (women of Liechtenstein) were only granted the right to vote in the 1984 referendum. Only men could vote. Only 51.3% were in favour

Danube

Hitler
Adolf Hitler was born here. He wanted to become a painter but was rejected by the Academy of Fine Arts Vienna and instead he devoted himself to a political career

BRAUNAU AM INN

Large Hadron Collider
Lying in a tunnel 27 km in circumference, it is the most complex experimental facility and the largest single machine in the world, built in 2009 for 7.5 billion euro

Its purpose is to investigate areas of physics for which current theories are inconclusive, such as the relationship between quantum mechanics and general relativity or the nature of elementary particles

Crime
Liechtenstein has one of the lowest crime rates in the world

Eagle's Nest
Hitler's headquarters from which he led the most important military operations

Salzburg Festspiel
The largest opera an theatre festival in Europ

○ SALZBURG

LIECHTENSTEIN

Only 32% of Austria's land lies below 500 m

ZÜRICH
The largest church clock face in Europe (8.64 m in diameter)

● BERN

SWITZERLAND

VADUZ
Population of 5,429

AUSTRIA

Wengen

The longest and fastest (up to 161 km/h) downhill skiing race

Wolfgang Amadeus Mozar
Wrote his first compositions an performed in front of Europe royalty at the age of At 8 years old he create his first symphon

Jungfrau-Aletsch protected area
The largest glaciated area in western Eurasia, including Aletsch Glacier, the largest in the Alps

GENEVA
The Red Cross was founded here in 1863

St Moritz
The oldest and one of the most expensive ski resorts in the world

Mozartkugel
Famous small balls of marzipan coated in a praline cream and dipped in warm chocolate

The Gotthard Base Tunnel

World's longest rail tunnel at 57 km. It took 17 years and $12 billion to build and 8 people lost their lives during construction

Matterhorn 4,478 m
Its characteristic shape and beautiful surroundings make it the most photographed mountain on the planet

The Alpine Club
During the golden decade, the world's first mountaineering club was founded in 1857 in London

Amber Road

MONTREUX
Known for the 2nd largest jazz festival in the world, and the casino fire in 1971. The incident was immortalized by the famous song "Smoke on the Water" by Deep Purple

Alps
The Alps include 128 peaks exceeding 4,000 m. Most of them are in Switzerland

The golden age of alpinism

Decade when most major Alpine peaks were first climbed successfully. It started in1854 with Wills's ascent of the Wetterhorn, and ended in 1865 with Whymper's ascent of the Matterhorn

0 — 50 — 100 — 200 km
0 — 25 — 50 — 100 miles

Gregor Mendel

Was a scientist and Augustinian friar known for establishing the science of modern genetics. He was born in the small village of Hynčice in 1822

Beer nation

With over 140 litres drunk per year per capita, Czechia has the world's highest consumption of beer

Austro-Hungarian Empire

For 51 years, Austria and Hungary formed the second largest empire after Russia. It collapsed in 1918, as a result of World War I

Prague Castle

It's the largest castle complex in the world

PRAGUE
Charles University in Prague, founded in 1348, is the oldest university in Central and Eastern Europe

Atheism

Up to 80% of burials are held without any ceremony and many burial urns are destroyed after being unclaimed by family members

Count Maurice de Benyovszky (1746–86)

Slovakian national hero, explorer & writer. Was exiled to Kamchatka by the Russians for opposing the partition of Poland

Stole a Russian ship and escaped. He travelled to the Aleutian Islands, Alaska, then to Japan, Taiwan and Macao

He tried to colonize Madagascar, where the local tribes made him a king. He was killed on the island by the French for supporting the American Revolution

CZECHIA

Budweiser Budvar

Established in České Budějovice in 1785. Since 1907, it has been in a legal dispute with American brewer AB InBev over the rights to the name

Battle of Austerlitz

Probably the greatest victory of Napoleon, over the larger armies of Russia and Austria (1805)

Czechoslovakia

Between 1918 and 1993, these two countries formed a single state

The country is similar in size to the Dominican Republic

Castle capital

Highest number of castles & chateaux per capita

Štefan Banič

Invented and patented an early design of the parachute

SLOVAKIA

Neutrality

In 1955, Austria declared itself permanently neutral

Venus of Moravany

Small figurine made of mammoth tusk, depicting a woman's body. It is over 24,800 years old, and was discovered by a farmer in the 1930s

Ján Bahýľ

Designed, built and successfully flew the first motor-driven helicopter

Invented by Hungarian expats:

Holography (Dennis Gabor)
The BASIC programming language (John Kemeny, Thomas Kurtz)
The ballpoint pen (László Bíró)
The theory of the hydrogen bomb (Edward Teller)

VIENNA
Home to the world's oldest zoo, established in 1752

BRATISLAVA

1986 Hungarian Grand Prix

It was the first F1 race behind the Iron Curtain, won by Ayrton Senna

Battle of Vienna

Won in 1683 thanks to the support of the Polish-Lithuanian army, it broke the advance of the Ottoman Empire, the furthest military advance of any Muslim Empire into Europe

The battle included the largest recorded cavalry charge

When the Turks fled after the defeat, they left large amounts of coffee beans, which initiated the tradition of the Kaffeehaus in Vienna

Danube

English

The lowest rate of English-speakers of all EU countries at around 20%

BUDAPEST
Before 1873 it was two separate cities – Buda and Pest. Today it's famous for its thermal baths and the world's largest underground thermal cave system

Lake Hévíz

World's largest thermal lake in which people can bathe (47,500 m²)

HUNGARY

Rubik's cube

It was invented by an architecture professor Ernő Rubik in 1974

Alpine skiing

The most successful country in alpine skiing (114 Olympic medals, including 34 golds)

Lake Balaton

Although it's the largest lake in Central Europe, it is very shallow, on average just 3.2 m deep

VAT

Hungary has the highest value-added tax in the world: 27% of sales net value

Marcus Aurelius

The last of the "Five Good Roman Emperors", died in AD 180 in modern-day Vienna during the war with Germanic tribes

His wrote "Meditations", the most comprehensive piece on ancient Stoic philosophy

Sigmund Freud

Austrian neurologist, founder of psychoanalysis. He shaped modern psychology by researching and redefining the role of sexuality. Because of his Jewish descent he fled from Vienna to London where he died in 1939

Amber Road

Danube

Yugoslavia
From 1918 to 1991, Macedonia, Slovenia, Montenegro, Croatia, Bosnia & Herzegovina, Kosovo and Serbia formed one country

Yugoslavia was at first a kingdom ruled by the Karađorđević dynasty, then in 1946 it became a socialist republic led by dictator Josip Tito

Lepenski Vir
Archaeological site which provides evidence for the earliest examples of monumental sacral art in Europe

Raspberries
Majority of the raspberries sold on the world market are produced here

Fruit rich
Serbia is one of the largest suppliers of frozen fruits to the EU

SERBIA

BELGRADE
The main street of today's Belgrade is part of the historic road to Istanbul

Pan-Serbism
Ideology that Serbs from all the Balkan states should unite territories to form a Greater Serbia; as it was before the Ottoman conquest

Nostalgia for the Serbian Empire was one of the reasons for the 1990s Balkan War

Yugoslavian basketball
With 5 titles, Yugoslavia is the most successful team in the FIBA Basketball World Cup (tied with the US)

European migrant crisis
Main route travelled by migrants through Europe during the crisis of 2015–16. 46.7% of the refugees were Syrian

Migrant route

Podgorica-Belgrade train route
Scenic route that crosses over 200 bridges and runs through more than 400 tunnels

Hungarian border barrier
In 2015, Hungary built a 523-km-long fence to prevent migrants from entering the country illegally

Hungarian border barrier

Croatian War of Independence
The war was fought between the Serbian army of Krajina and Croatia, which declared independence from Yugoslavia. It lasted 5 years from 1991

Srebrenica massacre
This 1995 massacre of over 8,000 Muslim Bosniaks was described as the worst case of genocide in Europe since WWII

Water security
Slovenia is rich in water resources and the right to drinking water is secured in the country's constitution

Independence
In June 1991, Slovenia was the 1st of the 6 republics to break away from Yugoslavia

Mountains
Over 70% of Slovenia's area is covered by the eastern Alps

Cravat
Modern neckties originated here around the 17th century and were spread by Croatian mercenaries enlisted by the French

CROATIA

Perućica
One of the last primeval forests in Europe. You can explore it only in the company of guides

BOSNIA & HERZEGOVINA

First trams
Sarajevo was the 1st city in Europe and 2nd in the world, after San Francisco, to introduce a fully electric tram network

SARAJEVO
Assassination of Archduke Franz Ferdinand of Austria in Sarajevo (1914) triggered the start of WWI

Ski jumping in Planica
One of the largest ski jumping hills with the highest number of world records

Jamica Kostelić
Has won 4 Olympic gold medals in alpine skiing, more than any other woman

KRK
The most populous Croatian island (19,383 people)

SMILJAN
Nikola Tesla was born here in 1856. He contributed to the invention of the alternating current electricity supply system, still used today

Winter Olympics
Yugoslavia hosted the Winter Olympics (1984) in the Sarajevo region, the only country from the Soviet bloc to do so

Migrant route

Dalmatians
Dalmatian dogs come from the historical southern region of Dalmatia

White House
The marble used to build the White House, comes from the Croatian island of Brač

SLOVENIA

LJUBLJANA

ZAGREB

Triglav National Park
The only national park in Slovenia

Coastline
Slovenia has a tiny (46 km long) but crucial sea access

Istria
The largest peninsula on the Adriatic

CRES
The largest Croatian island (405.78 km²)

Islands
Croatia has 1,246 islands with an area of around 3,300 km², corresponding to the area of northern Cyprus

Wars

In recent decades Serbia has fought with:
Croatia (1991–95)
Bosnia & Herzegovina (1992–99)
Kosovo Albanians (1998–99)
Albanian separatists (1999–2001)

Mother Teresa

Agnes Gonxha Bojaxhiu was born in Skopje in 1910 in the former Ottoman Empire

MACEDONIA

SKOPJE

Alexander III of Macedon

The empire of Alexander the Great covered 3.5% of the world's land area in 323 BC

Arbor Day

After the wildfires of 2007, Macedonians took action to restore their forests and planted over 6 million trees in a single day. Since then it's become an annual tradition

Etymology of the word "Balkan"

In Turkish it means "wooded mountain chain"

The word "Balkans" was first used to describe the region on a 14th-century Arab map

Since the Late Middle Ages, the region's culture has been influenced by the Ottoman Empire

Mountains

The majority of the region's land area is covered by mountain ranges

Kosovo adopted the euro as its national currency although it's not part of the eurozone

PRISTINA

KOSOVO

MONTENEGRO

PODGORICA

Tara River

Canyon

The deepest canyon in Europe (1,300 m)

Bosnia's coastline is only 20 km long

DUBROVNIK

Beautiful city, listed as a UNESCO World Heritage site, is one of the most popular tourist destinations in the Mediterranean

Marko Miljanov

This great Montenegrin general and writer learnt to write only when he was 50

Oldest olive tree

The Old Olive of Mirovica is one of the oldest trees in the world (over 2,000 years old)

Bojana river

The only river in the world that occasionally runs upstream. This may happen during winter when the waters of its tributary cut across and push part of the river flow upstream

ADRIATIC SEA

The salinity of this sea is lower than the Mediterranean because around 1/3 of the fresh water flowing into the Mediterranean, flows into the Adriatic first

King Zog

There were over 50 assassination attempts on the life of Albania's King Zog

ALBANIA

TIRANA

Bunkers

Albanian dictator Enver Hoxha, in fear of a Soviet invasion, ordered the construction of 750,000 concrete bunkers

No cars

Until 1991, private car ownership was banned here with a rate of 2 vehicles per 1000 people

Albanian Mafia

The country has an infamous reputation for organized crime, especially drugs and human trafficing

Strait of Otranto

Width of 72 km

Otranto Barrage

An allied naval blockade of the strait restricted the Austro-Hungarian navy to the Adriatic Sea during World War I

ADRIATIC SEA

KOSOVO

Ethnic war

Suffered from a violent ethnic conflict with Serbia during the late 1990s. Despite NATO intervention about 10,000 people were killed and 1 million lost their homes. Kosovo declared independence in 2008, but Serbia and Russia refused to recognize it

Islam

It is one of two Muslim majority states in mainland Europe and at the same time ranked as one of the most religiously tolerant states in the world

Size, population & economy

Kosovo is the size of Jamaica, it has a population of just 1.8 million people and has the 5th largest lignite reserves in the world

MEDITERRANEAN SEA

200 km

100 miles

100

50

50

25

0

0

First historian
Herodotus is the first known historian. In the 5th century BC, he wrote the history of Greece

First ice cream
In the 5th century BC, ancient Greeks ate a dessert consisting of mixed snow, fruit and honey

Dark ages
After the collapse of Mycenaean civilization around 1100 BC, Greece entered into 300 years of the "dark ages" and lost the knowledge of writing

Anaximander's world map
First attempt to illustrate the entire known world at that time (6th century BC)

1st Ancient Olympics
776 BC
Several city-states
One event – running race

1st Modern Olympics
1896
14 nations, 241 athletes
43 events

Coins
Alexander the Great became the first ruler of Greece to have his image placed on coins – before this they portrayed only the images of gods

Alexander the Great
Undefeated in battle, created an empire stretching from Greece to India, and died in unclear circumstances in Babylon

Heritage
Greece is the birthplace of democracy, Western philosophy and literature, political science, major mathematical principles, and tragedy and comedy drama

Unemployment
In 2015, the youth unemployment rate in Greece was 49%, the highest in the EU

PELLA
Birthplace of Alexander the Great in 356 BC

THESSALONIKI
Known as the "co-capital" of Greece

Drachma
Europe's oldest currency, established over 2,650 years ago and replaced in 2002 by the euro

Rivers
Greece has no navigable rivers because of the mountainous terrain

Mount Olympus △
The tallest mountain in Greece (2,911 m). In Greek mythology it was home to the twelve Olympian gods

GREECE
Its official name is Hellenic Republic

AEGEAN ISLANDS
Divided into 7 main groups: Sporades, North Aegean, Dodecanese, Cyclades, Saronic, Crete, Evvoia

The territory of ancient Greece, for many years, included the west coast of present-day Turkey

Coastline
Greece has over 13,000 km of coastline, 11th longest in the world

Olympic sacrifice
Greeks used to sacrifice 100 bulls to Zeus during the Olympics

SPORADES

NORTH AEGEAN ISLANDS

EVVOIA

AEGEAN SEA

Aristarchus of Samos
The first man who proposed the theory that Earth revolves around the Sun. It was popularized 1,700 years later by Nicolaus Copernicus

Marathon
The original marathon route run by the Greek messenger Pheidippides in 490 BC to announce the news of the victory of Athens in the Battle of Marathon

Cradle of democracy
Cleisthenes reformed the Athenian constitution and introduced the foundations of the world's first democracy in 508–507 BC

ATHENS

SARONIC ISLANDS

CYCLADES

★ **Thales of Miletus**
First philosopher (624–546 BC). The first one to abandon mythology as a basis for explaining the universe and to use science and logic instead

Ancient Olympia
Ancient Olympic Games site

DODECANESE

Colossus of Rhodes
The tallest statue of the ancient world and one of its Seven Wonders. It was about the size of the Statue of Liberty. It was destroyed during an earthquake in 226 BC

RHODES

Calypso Deep ★
The deepest point of the Mediterranean Sea, with a depth of 5,267 m

Ancient Sparta
Spartan military training started at the age of 5

Greek islands
Greece has up to 6,000 islands and islets, but only 1,200 are habitable and only 227 are permanently populated

Knossos
Location of the Cretan Palace and Minotaur's Labyrinth

IONIAN ISLANDS

IONIAN SEA

First civilization
Minoan civilization of Crete is the oldest recorded European civilization (3650–1400 BC)

CRETE

IRAKLION
Ottoman forces besieged Iraklion for 21 years. It was one of the longest sieges in history

Crete is the geographical crossroads between three continents

Monk seal
The only species of earless seal which inhabits sub-tropical waters. It's extremely rare with just a few hundred surviving in the wild

Greek Merchant Navy
Since ancient times Greece has relied economically on maritime trade. Its merchant fleet accounts for over 16% of the world's total, making it the largest in Europe and one of the five largest in the world

Icarus
Icarus tried to escape from Crete when he flew too close to the sun and the wax holding his wings together melted

Zeus
Mythological Greek king of the gods was born in Crete. According to a legend, Heracles removed all dangerous animals from the island to honour his father Zeus

0	50	100		200 km

0	25	50		100 miles

Warsaw Pact

Military pact established in opposition to NATO in1955 by the Soviet Union and its 7 central-eastern European satellite states. The pact failed in 1989 after the violent revolution in Romania and killing of the dictator, Nicolae Ceaușescu. The treaty was officially disbanded 2 years later in Prague

Nazi ally

Before WWII, Romania was an ally of France and Britain. However, with the fall of France, the government decided to join Nazi Germany in 1940

Romanian treasure

During WWI, the Romanian government handed over to Russia around 120 tonnes of gold for safekeeping. To this day the country has recovered almost none of this

Traian Popovici

Romanian lawyer who saved approximately 20,000 Jews during the Holocaust

Home ownership

96% of Romanians own their homes (highest in the world)

Romance language

Romanian is the only Romance language in Eastern Europe, closely related to languages such as French and Italian

Brown bears

Romania has the world's largest population of brown bears, around 6,000

Turda Salt Mine Park ★

In antiquity, one of the largest salt mines in the world. Today it's the deepest located amusement park featuring a ferris wheel and boating in the kaleidoscopic cave lake

Gold

Europe's richest country in gold

Hungarians

Over 80% of the population in the county of Harghita is Hungarian

Transylvania

Beautiful, historical region of Romania, rich in minerals, commonly known as the land of vampires

Nadia Comăneci

The first gymnast to be awarded the perfect score of 10 at the 1976 Olympics in Montreal. She was just 14 years old back then

Danube Delta

UNESCO World Heritage site, one of the wildest places in Europe. The river carries 2 tonnes of silt a day into the Black Sea

Carpathian Mts

ROMANIA

Bran Castle ★

Known as the legendary "Dracula's Castle"

The centre of the country lies halfway between the North Pole and the Equator

Letea Forest

The northernmost sub-tropical forest in the world

TIMIȘOARA

TIMIȘOARA

The first city in Europe and the second one in the world (after New York) to introduce electric street lighting, in 1884

Transylvanian Alps

Danube

Dacia

Dacia is a Romanian car company. In 2015 it sold over 100,000 of its cars in France. Its headquarters are located in the small town of Mioveni

Transfăgărășan

Scenic road, full of sharp turns and steep descents, named as greatest road in the world by the host of the famous Top Gear TV show

Danube

BUCHAREST

Europe's Mt Rushmore

The sculpture of king Decebalus is located in the scenic Iron Gates canyon on the Danube river. It's the largest rock sculpture in Europe, and more than twice the size of the Mount Rushmore sculptures

Romanian Palace of Parliament

2nd largest administrative building in the world, after the Pentagon. It has 1,100 rooms

BLACK SEA

Danube

2,850 km long, it is the 2nd longest river in Europe (after the Volga). It runs through 10 countries, more than any other river in the world

Yogurt

Bulgarian yogurt in considered the best in the world thanks to a unique bacteria used in it

Roma people

Widely known as "Gypsies" they are a nomadic ethnic group that originated in India. Their total population is estimated to be 2–20 million people

Christianity

98% of the population is Christian (4th in the world)

VARNA

VARNA

Summer capital of Bulgaria

Battle of Varna

Great Ottoman win over the Hungarian-Polish Crusade army (1444), not supported by a papal and Venetian fleet, despite promises. Resulted in the death of King Władysław III of Poland and Hungary

BULGARIA

SOFIA

SOFIA

Roman Emperor Constantine the Great (AD 282–337), considered moving the capital of the empire to here

World's oldest gold jewellery dating back to 4600–4200 BC was found in Varna

Rose Valley ★

Region famous for its rose-growing industry, accountable for 85% of the world's rose oil production

Rafail's Cross

It's a well know crucifix, located at Rila Monastery, made of one piece of wood. It depicts over 100 religious scenes including 650 miniature figurines. The monk who made it lost his sight after 12 years of making the piece

PLOVDIV

PLOVDIV

2nd most populous city in the country (341,567 people). It is also the 3rd oldest continuously inhabited city in Europe (since 4000–3000 BC)

Bulgarian border barrier

Built along the Bulgarian-Turkish border as a response to the European migrant crisis. The construction started in 2014

The Cyrillic alphabet was invented in Bulgaria in the 9th century. After the accession of Bulgaria to the EU, the Cyrillic alphabet became its 3rd official alphabet

| 50 | 100 | | 200 km |
| 25 | 50 | | 100 miles |

Dniepe

UEFA Euro 2012

Ukraine hosted jointly with Poland. It was the first time a post-Soviet country organized an event of this rank. Spain beat Italy 4–0 in the final

Once Polish lands

Eastern regions of Ukraine are the area lost by Poland after WWII. In return, it gained German lands to the west

Fertile lands

Around 2/3 of the country's land is covered with fertile "black earth"

Wildlife

The abandoned zone around the Chernobyl nuclear power station is rich in wildlife as animals here do not have to compete with man. However, there are many cases of mutation of the fauna

CHERNOBYL

The 1986 Chernobyl nuclear accident resulted in the release of 400 times more radioactive material than at Hiroshima. Almost all of Europe was affected

7 million people lived in the areas that were contaminated

Giant crossword

World's largest crossword puzzle, over 30 m tall, is painted on the wall of one of the residential buildings in Lviv

Demographic crisis

Since the 1980s, the population of Ukraine has been shrinking by over 150,000 people a year

KIEV

The deepest metro station on the planet can be found in the capital of Ukraine (>100 m deep)

LVIV o

Home to the first gas lamp ever made, by a pharmacist in a store called "At the Golden Star"

Space race

Ukrainian Sergei Korolev was a leading rocket engineer for the Soviet Union during the space race against the US

Battle of Kiev

In 1941, Germans captured the city. 600,000 Ukrainians were killed during the battle. Kiev was awarded the title "Hero City" for its heroic struggle

Eternity restaurant

Restaurant in a giant, 20-m-long coffin

Space flights

After gaining independence, only one Ukrainian has travelled to space, in 1997

UK

Priest's Grotto ★

Giant system of caves which served as a refuge for Jews during WWII. Some of these refugees did not leave the caves for 344 days

Kamianets-Podilskyi Castle

Known as "urbs antemurale christianitatis" (Bulwark of Christianity) this stronghold defended Eastern Europe from the Turkish and Tartar invasions for hundreds of years

World War I

During WWI Ukrainians had to fight against each other as soldie of the two enemy armies – Austro-Hungarian and Russian

Carpathian Mts

Carpathian Mountains

1,500 km in length, it is the 2nd longest mountain range in Europe (the Scandinavian mountain range is 1,700 km long). It partially covers several countries: Czechia, Slovakia, Poland, Hungary, Ukraine, Romania and Serbia

Pip Ivan Observatory

Beautiful meteorological observatory built high in the mountains in 1938. It operated only until the start of WWII

Moldova river

The name of the country comes from the Moldova river which flows through modern-day Romania before joining the Siret river

Transnistria

An autonomous republic recognized only by Russia and legally part of Moldova. Transnistria has its own currency, president and government

Strategic Missile Forces Museum

Old nuclear base transformed into a museum. You can see there, among other things, nuclear missiles, which were to be installed on Cuba during the Cuban Missile Crisis in 1962

Moldova

Wine

Grape wine represents 5% of Moldova's exports

MOLDOVA

Siret

Military expenditure

Growth of Ukraine's military expenditure (as a percentage of GDP) since the beginning of the armed conflict with Russia in 2014

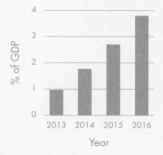

Mileștii Mici

World's largest wine cellars extending for 200 km

● CHIŞINĂU

After gaining independence from the Soviet Union in 1991 a large-scale operation to rename the capital's communist street names was carried out

Odessa was awarded the title "Hero City" for its defence during Operation Barbarossa in 194

ODESS

Deportations

On Stalin's orders, during two days in July 1949, 35,796 Moldovans were deported to Siberia to destroy potential anti-Soviet opposition

Odessa catacombs

The largest catacomb system in the world, around 2,500 km long

Gagauz people

Turkic-speaking group of approximately 240,000 people living mostly in southern Moldova. Throughout their history they have been an independent nation for only 5 days, in 1906

0   50   100   200 km

0   25   50   100 miles

BLACK

SEA

NATO membership
With the Russian invasion of 2014, it is now a priority for the Ukrainian government to join NATO

Leonid Brezhnev
This General Secretary of the Communist Party of the Soviet Union was Ukrainian. He served for 18 years, second only to Stalin

Population of Ukraine
Others 5%
Russians 17%
Ukrainians 78%

SLAVUTYCH
brand new city built after the Chernobyl disaster to resettle residents

Language
The Ukrainian language is very similar to Russian. It used to be known in Russia as the "Little-Russian" dialect

Nuclear arsenal
Ukraine could have been the 3rd largest nuclear power in the world, but its remaining nuclear warheads were given to Russia after gaining independence

"Holodomor"
Famine in Soviet Ukraine (1932–3) that killed between 2.4 and 7.5 million people. It was planned by Joseph Stalin to destroy the spirit of nationalism in the nation

Military
Ukraine has the 3rd largest military force in Europe (250,000 active personnel) after Russia and Turkey

KHARKIV
2nd most populous city in Ukraine (1.4 million people), capital of the Ukrainian Soviet Socialist Republic from 1919 to 1934

Donetsk People's Republic and Luhansk People's Republic
Two self-proclaimed states founded by separatist forces, financially and militarily supported by Russia. The armed conflict in these areas has been ongoing since 2014

Dnieper

Kremenchuk Reservoir

"Automagistrals"
There is only one motorway in Ukraine of considerable length ~180 km

Corruption
Ukraine is the most corrupt country in Europe and is ranked 136th in the world next to countries like Paraguay and Comoros

Pornography
Pornography is illegal in Ukraine

AINE

Largest country situated entirely within the European continent. 46th in the world

Similar size to Madagascar and South Sudan

Malaysian Boeing catastrophe
In 2014, pro-Russian insurgents mistakenly shot down a Malaysia Airlines Boeing resulting in 298 deaths

Dnieper reservoirs
Large artificial lakes created after construction of hydroelectric dams

DNIPROPETROVSK
One of the main centres of Soviet nuclear and space research

"The" Ukraine
In English, Ukraine was always referred to as "The Ukraine". After independence in 1991 "the" has been deleted from its name

Europe's breadbasket
Ukraine is one of the world's largest grain producers due to its large area of fertile land

Dnieper Hydroelectric Station
The largest hydroelectric power station in Ukraine. It was built in 1932 and heavily damaged in 1941 by the retreating Soviet army after the German invasion. It was damaged again by German troops in 1943 when they were retreating

DONETSK
Mining and steel production centre of the country

Kakhovka Reservoir

Viktor Yushchenko
Pro-European president, in power from 2005 to 2010. During his election campaign in 2004 there was an assassination attempt on his life with the use of poison

Zaporizhia Nuclear Power Plant
By capacity, it is the largest nuclear power plant in Europe and 5th in the world

Dnieper

Dnieper
4th longest river in Europe. The river is economically important for Ukraine because of its hydroelectric stations and water transport of goods

SEA OF AZOV
The shallowest sea in the world with an average depth of 7 m. Its maximum depth is only 14 m

First settlements
The first settlements in the territory which is now Ukraine were in the Crimea around 32,000 BC

Swamp
Ancient Greeks called the sea "Maeotian Swamp"

Crimea
Beautiful region that has always been the main tourist destination for Ukraine and the surrounding areas. In 1954 it became a gift from Stalin to Ukraine

Underwater Museum
Museum showing the collection of statues of communist leaders sunk several metres under water

Greek colonies
Traces of ancient Greek and Roman colonies have been found on the northeastern shore of the Black Sea

Crimea annexation
After the unrest caused by the 2014 Ukrainian revolution, Russia annexed Crimea and continues to administer the area.

Yalta Conference
Meeting of Stalin, Roosevelt and Churchill (1945), to discuss the future of Europe after WWII

Key points included splitting Germany into 4 zones, war reparations, creation of a communist block and establishing a socialist government in Poland

SEVASTOPOL
Even after the collapse of the Soviet Union it remained the main port of the Russian Black Sea Fleet

World's largest individual flower

Rafflesia arnoldii weighs around 7 kg and can
grow up to 105 cm in diameter. When it blooms,
it releases an odour comparable to rotten meat.
It is endemic to Sumatra and Borneo

ASIA

ASIA

Area
Asia comprises around 9% of the planet's area and 30% of the Earth's land surface with an area of 45,036,492 km²

$
Wealth
Asia has the largest economy of all the continents. Its GDP is worth $27.2 trillion which constitutes 36% of the world's total.

Religions
All four of the largest religions, or spiritual systems, of today originated in Asia. Altogether, their followers make up over 70% Earth's population

Highest and lowest
Both of these extreme points of the Earth, Mount Everest and the Dead Sea, are located in Asia

Tallest buildings
Of the 10 tallest buildings in the world, 9 are located in Asia and 5 of these are in China. The only exception is the One World Trade Center in New York

48 states
Asia is composed of 48 sovereign states, including Turkey and Russia which extend in to Europe

Silk Road
Ancient trading and cultural network of routes extending 10,000 km, which linked the regions of the ancient world. Its name comes from the popularity of Chinese silk in the West (especially in Rome)

Cape Baba
Westernmost point in mainland Asia 26°04'E

Wheel ★
The first wheeled vehicle might have been used in what is now Syria sometime between 6500 and 5100 BC

Christianity ✝
Jesus Christ, the religion's central figure, was born in Bethlehem. Today, Christianity has 2.2 billion followers – 31.5% of the world's population

★ Islam ☪
Prophet Muhammad, the founder of Islam, was born in Mecca around AD 570. Today, about 22% of the world's population, or 1.6 billion people, are Muslim

Black Death ☠
Originated in Central Asia from where it travelled with the Mongol army to the Crimea where Mongols besieged the city of Kaffa. The army suffered from the disease and catapulted infected corpses over the walls of the city. The inhabitants fled to Sicily where the plague spread

Early Muslim conquests ☪
The territory of the Umayyad Caliphate (AD 661–750) stretched from the Iberian peninsula to the borders of China

Yoga
Originated during the Indus Valley Civilization 3300–1900 BC

First Nobel Prize
Rabindranath Tagore, an Indian polymath, was the first Asian to win a Nobel prize (for literature in 1913)

ARCTIC OCEAN

Gulf of Ob
The estuary of the Ob river is the longest in the world

Ob-Irtysh
Westernmost of the great Siberian rivers, 2nd longest in Asia and 5th longest river system on the planet (5,568 km)

Eurasia
There is no clear physical separation between Asia and Europe

Reinhold Messner 🚶
In 1986, this Italian mountaineer became the first man to climb all the 8,000-ers. He was also the first to cross Antarctica and Greenland entirely on foot

Himalaya

DELHI
Population: 27.2 million

Buddhism
Buddha was born in what is now Nepal around 560 BC. His disciples do not regard him as a god, but rather as an enlightened teacher, guiding buddhists on the path to liberation from suffering in life

Hinduism
Considered the oldest system of spiritual practices and beliefs. It has no founder, includes millions of deities and its roots reach beyond any written accounts. It's followed by 1 billion people, 14% of the world's population

Ganges Delta
World's largest river delta and one of the most fertile regions. Despite cyclical floods and poverty, it supports the lives of 300 million people

Largest agglomerations
Of the 10 largest urban agglomerations on the planet, 7 are located in Asia: Tokyo, Delhi, Shanghai, Beijing, Mumbai, Osaka and Dhaka

Richest countries in Asia
$ 6 richest countries in Asia by nominal GDP compared with nominal GDP per capita

Country	Nominal GDP (millions of US$)	Nominal GDP per capita (US$)
China	11,218,281	8,113
Japan	4,938,644	38,917
India	2,256,397	1,723
South Korea	1,411,246	27,539
Russia	1,280,731	8,929
Indonesia	932,448	3,604

INDIAN OCEAN

0	500	1,000	2,000 km
0	300	600	1,200 miles

Cape Chelyuskin
Northernmost point in mainland Asia 77°43'N

Cape Dezhnev
Easternmost point in mainland Asia 169°40'W

Russian Empire
At its greatest extent in 1895, it occupied 15.31% of the world's land area

Yenisey–Angara–Selenga
This 5,550-km-long river system is the 3rd longest in Asia and 6th longest in the world

Eurasian Steppe
Mostly grassland plains, stretching from east of the Danube basin to almost the Pacific Ocean. It played an important role in the spread of cultures, tools, languages and skills

The Long Walk
In 1941, 7 POWs escaped a Soviet labour camp in Siberia and trekked over 6,500 km to India. A book "The Long Walk" and a Hollywood movie "The Way Back" were based on this unconfirmed history

The Long Walk

Cradle of civilization
The first great civilizations on Earth began in Asia:

| China 7000 BC | Indus Valley Civilization 7600 BC | Fertile Crescent 10200 BC |

Diversity
Asia is the most diverse continent in terms of fauna, flora and climate

BEIJING ●
Population: 22.1 million

Longest pipeline

Chang Jiang (Yangtze)
The longest river in Asia and the 3rd longest in the world (6,380 km)

Chang Jiang (Yangtze)

SHANGHAI
Population: 25.2 million

Longest pipeline
West–East Gas Pipeline III, with its 5,220-km trunkline and 8 branches, is 7,378 km long

DHAKA
Today it has a population of 18.9 million people. By 2075, Dhaka is expected to become the largest city with a population of 57 million

Southeast Asia
Subregion of Asia consisting of: Vietnam, Laos, Cambodia, Thailand, Myanmar, Malaysia, Indonesia, Singapore, Philippines, East Timor and Brunei

Four Asian Dragons
Taiwan, Singapore, Hong Kong (China) and South Korea are high-tech industrialized countries that underwent rapid development between the 1950s and 1990s

Tanjong Piai
Southernmost point in mainland Asia 1°14'N

Toba supereruption
One of the largest known volcanic eruptions (about 75,000 years ago), caused a 6-year-long global winter

Population
Asia is the most populous continent on the planet, and it is likely to remain that way until the end of the 21st century. More people live in Asia than in all the other continents combined

Asia (59.69%)
4,436,224,000 people

Sakoku
Isolationist foreign policy of Japan from 1633 to 1853. During this period, trade between Japan and other countries amounted to almost nothing

TOKYO
Population: 38.2 million

Japanese Empire
By 1942, the Japanese Empire encompassed the territories of Manchuria, East China, Southeast Asia and many of the Pacific islands

"The Travels of Marco"
Marco Polo's journey to Asia inspired Columbus to explore the world and influenced the creation of the Fra Mauro map, which marked the end of the Bible-based approach in the history of cartography

PACIFIC OCEAN

Megadiverse countries
5 of the 17 megadiverse countries (a group of countries that are home to the majority of species on Earth) are in Asia: China, India, Indonesia, Malaysia and the Philippines

Islamization
Islam first came to Southeast Asia with Muslim traders. The Sultan of Kedah (Malay Peninsula) was the first ruler of this region to convert to Islam in 1136

"Third World"
Many think that this term refers to the level of a country's development, but it was originally used to describe countries unaligned with either the Soviet Bloc or NATO during the Cold War

ARCTIC
OCEAN

Franz Josef Land
This archipelago, named after the Emperor of Austria and King of Hungary Franz Joseph I ,was discovered in 1873. It is inhabited exclusively by Russian military personnel

FRANZ JOSEF LAND

BARENTS SEA

Kaliningrad Oblast
Originally part of East Prussia. It was annexed by the Soviet Union after the defeat of Nazi Germany
Economically it is the best performing region in Russia

MURMANSK
Despite its northern location, around 300,000 people live here. It has highway and railway access to the rest of Europe and the northernmost trolleybus system on Earth

*NOVAYA
ZEMLYA*

KARA SEA

Lake Ladoga
The largest lake in Europe, 14th largest in the world

ST PETERSBURG

Lake Onega
2nd largest lake in Europe

AK47
Invented and produced in the USSR, it is the most widely used type of rifle in the world. Around 100 million units have been produced

Yenisey

Smolensk catastrophe
In 2010, a Polish aircraft crashed, killing all passengers on board including the Polish president and his wife, members of parliament and senior officers of the armed forces

*Ural
Mountains*
They divide Russia into its European and Asian parts. The Urals are one of the world's oldest mountain ranges (250 – 300 million years old)

Yenisey-Angara-Selenga river
World's 6th longest and the largest river system that flows into the Arctic Ocean

MOSCOW
One of the world's most expensive cities. In 2014, 85 billionaires were living in Moscow

$

Trans-Siberian Railway

Ob

West Siberian Plain
The largest plain in the world

Illegal flight
In 1987, a young German Mathias Rust, piloting a single-engine plane, flew from Helsinki and landed in Red Square in Moscow in order to symbolically reduce the tension of the Cold War

*Virgin
Komi Forests*
The largest primeval forest in Europe

Ob

Size
The total area of Russia accounts for 11% of land on Earth and is similar in area to Pluto. It extends across 11 time zones

Women
There are approximately 10 million more women than men in Russia

*Russia's
European section*
Russia's European section is home to most of Russia's industrial and agricultural activities

Radioactive lake
The Soviet Union used Lake Karachay to dump nuclear waste. It has been described as the most polluted spot on the planet. An hour in its vicinity may result in death

Irtysh

Ob–Irtysh river
World's 5th longest river. One of the three great Siberian rivers along with the Yenisey and Lena

VOLGOGRAD
Formerly Stalingrad (1925–61). Historically an important industrial centre

Ural Mountains

Caucasus

Battle of Stalingrad
Fought between 1942–3, it is often described as the largest and bloodiest battle in history. Won by the Soviets, it was a turning point for WWII and resulted in an estimated 2 million casualties

Altai Mountain

Communism
The origins of this ideology can be traced back to Ancient Greece and "The Republic" by Plato. "The Communist Manifesto" (1848) was written by Germans Karl Marx and Friedrich Engels

Communist countries today: China, Cuba, Laos, Vietnam, North Korea

| 0 | 250 | 500 | | 1,000 km |
| 0 | 150 | 300 | | 600 miles |

CHUKCHI SEA

Bering Strait

BERING SEA

EAST SIBERIAN SEA

Sakha Republic
The biggest country subdivision in the world (3,083,523 km²), similar in size to India. With only 958,528 people living there, it has a population density of 0.31/km², lower than Mongolia, the least densely populated country

End of the Cold War
In 1963, the Soviet Union and the US were close to joining forces during the space race to land on the moon. Unfortunately, Nikita Khrushchev backed out after President Kennedy was assassinated

Permafrost
65% of Russian territory is covered in permafrost with methane, a strong greenhouse gas, trapped inside. As the Earth warms, the melting permafrost may release carbon in to the atmosphere which could have catastrophic consequences for the planet

Sense of wilderness
Prior to 1990, Kamchatka was a closed military enclave. Foreigners were not allowed access and even Russians needed permits to enter

Gulags
Labour camps during the Stalin era have been said to have held over 50 million people in prisons. Most were worked to death

Klyuchevskaya Stopka
The largest (4,688 m) and the most active volcano in Eurasia and the whole Northern Hemipshere

Nuclear arsenal
Russia is in possession of more warheads than any other country in the world. The Soviet Union has built about 55,000 nuclear warheads since 1949

NEW SIBERIA ISLANDS

EVERNAYA EMLYA

LAPTEV SEA

Kamchatka
Kamchatka has around 160 volcanoes of which 29 are still active

ake aymyr orld's largest ke north of the ctic Circle

Putorana Plateau
This nature reserve, a UNESCO World Heritage site, protects the largest herd of reindeer and snow sheep. It holds one of the largest nickel deposits on the planet, has around 25,000 lakes and has the tallest waterfall in Asia. Its ecosystem remains untouched by humans

Siberia
It makes up around 75% of Russia. Extraordinarily rich in minerals, containing sources of almost every valuable metal

"The bear kingdom"
One of the highest concentrations of bears in the world. Up to 30,000 inhabit Kamchatka

SEA OF OKHOTSK

Hydroenergy
Russia has well-developed hydropower stations. It is 5th in the world when it comes to the production of renewable energy, but only 56th when hydroelectric energy is not taken into account

Siberian tiger
The largest member of the cat family. Only about 540 remain in the wild

Lena

Lena river
World's 13th longest and the longest that flows entirely within Russia

YAKUTSK
The coldest (record low temperature of -64.4°C) large (around 326,457 people) city in the world. In winter the city is accessible only by the frozen Lena river highway

Still at war
Russia and Japan did not sign a peace treaty after World War II due to a dispute over the Kuril Islands

SAKHALIN

Mir mine ★
Former diamond mine, the largest excavated hole in the world, over 1 km in diameter and 0.5 km deep

Tunguska event
In 1908, a meteor hit Russia destroying 2,000 km² of forest, but no impact crater has been found. It is considered the largest impact event on Earth in recorded history

Stanovoy Mountains

Forests
Russia has the world's largest forest reserves, known as "the lungs of Europe". Second only to the Amazon rainforest in the amount of carbon dioxide absorbed

Mammoth
Remains of this extinct genus were first discovered in Siberia

KURIL ISLANDS

Trans-Siberian Railway

Lake Baikal
World's deepest (1,642 m) lake and largest freshwater lake by volume. Contains roughly 1/5 of the world's unfrozen fresh surface water

Sayan Mountains

Trans-Siberian Railway
The longest railway in the world (9,289 km), passing through 8 time zones. It takes 8 days to complete the journey

Building started in 1891 at both ends and worked towards the centre. Soldiers and convicts provided labour for its construction

SEA OF JAPAN (EAST SEA)

Yenisey

Baikal seal
The lake is the habitat of this endemic seal, one of the smallest and the only freshwater seal in the world

VLADIVOSTOK
Eastern Russia's only ice-free port. A base for whaling fleets and a terminal point of the Trans-Siberian Railway

Leonardo Bridge Project

In 1502, Leonardo da Vinci designed a bridge to run across the Golden Horn. It would have been the world's longest single masonry arch span today, however, Sultan Bayezid II did not approve the plans

A smaller version of this design was realized in 2001 in Norway. It was the first completed civil engineering project based on Leonardo's designs

Anoxic waters

The deep waters do not mix with the upper layers, rich in oxygen. As a result, most of the water at great depths is anoxic (lacking in oxygen), which keeps archaeological artifacts, such as ancient shipwrecks, in very good condition

BLACK SEA

One of 4 seas that has a colour in its name, the others being: Red Sea, White Sea and Yellow Sea

Great flood

There is a hypothesis that waters from the Mediterranean breached the natural sill in the Bosporus strait resulting in a catastrophic flood of the Black Sea, around 5600 BC

Burning water

The Black Sea occasionally seeps combustible gases which, during thunderstorms, results in observable isolated flares when lightning hits

European part

3% of Turkey's area
10% of the population

Asian part

97% of Turkey's area
90% of the population

Hagia Sophia

For almost 1,000 years, it was the largest Christian cathedral, until it was converted into an Ottoman mosque in AD 1453

Bosporus

The geographic border of the continent

SEA OF MARMARA

ISTANBUL
The largest city spanning two continents. It has been the capital of three empires:
Roman Empire
Byzantine Empire
Ottoman Empire

European Union

Turkey was a candidate for EU membership, and started the accession negotiations in 2005. The process was slow, however, and eventually came to a halt after the 2016–17 purges and human rights violations

Failed putsch and the aftermath

In July 2016, a faction of the Turkish military led an unsuccessful attempt to overthrow President Erdogan and public institutions in Turkey. Since then, tens of thousands have been arrested and hundreds of thousands have been laid off work on political grounds

Troy

The ancient city immortalized in Homer's "Iliad"

Geopolitical importance

Turkey's location on the border of two continents and its control over the "Turkish Straits" has granted it geopolitical and strategic importance

ANKARA
Ankara replaced Istanbul as the capital after the fall of the Ottoman Empire in 1922

TURKEY

The only considerably secular country in the Muslim world

Anatolia

It's the westernmost Asian peninsula and the region known as Asia Minor, one of the oldest permanently settled regions in the world since the Palaeolithic era

Agricultural production

Turkey is one of the countries which is self-sufficient in terms of food production. It is the world's largest producer of hazelnuts, cherries, figs and apricots and one of the largest producers of watermelons, cucumbers, chickpeas, tomatoes, aubergines, green peppers, lentils and pistachios

Cappadocia

Ancient region of Turkey known for its lunar landscape, and houses and churches carved in rock

Temple of Artemis

Dedicated to the goddess of animals. One of the "Seven Wonders of the Ancient World". Destroyed in AD 401

Military

Turkey has the 10th strongest military in the world. Every fit male must serve

Refugees

Turkey is currently the largest refugee host with over 3 million people who have fled from Syria and Iraq

Mausoleum at Halicarnassus

One of the "Seven Wonders of the Ancient World", it was a tomb built for a Persian ruler, Mausolus

BODRUM

Antioch earthquake

This AD 526 earthquake killed approximately 250,000 people

Saint Nicholas

This saint, known as "Santa Claus", was born here

Oldest shipwreck

Bodrum Museum of Underwater Archaeology has the oldest ever recovered shipwreck dated between the 13th and 14th centuries

Syrian
Civil War
1.1 million Syrians have fled the country since the outbreak of the civil war in 20...

British rule

Between 1878 and 1960, Cyprus was ruled by the British

Divided capital

Nicosia is divided between the two parts of Cyprus

NICOSIA

Division

The division between the Turkish Republic of Northern Cyprus, which is only recognised by Turkey, and the independent Republic of Cyprus in the south

Mass destruction

The Syrian government has used sarin, chlorine gas, cluster bombs and thermobaric weapons against opposition forces. Civilians have been the largest group of victims

Aphrodite

According to Greek mythology, the goddess of love was born on Cyprus

CYPRUS

Oldest wine

Commandaria, a sweet dessert wine, is considered the world's oldest named wine still in production (since the 12th century)

Halloumi cheese

Traditional Cypriot cheese, made from a mix of goat's and sheep's milk, perfect for grilling

Country on a flag

Apart from Kosovo, it is the only country to put its country's shape on its flag

First Olympic medal

In 2012, Cyprus won its first and only Olympic medal, silver in the men's laser sailing

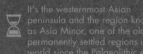
MEDITERRANEAN SEA

Ceasefire

A ceasefire line was established by the UN in 1974 after the Yom Kippur War fought between Israel and a coalition of Arab states

DAMASCUS
First inhabited around 10,000 BC. Its old town dates back to the 3rd millennium BC. It was also the capital of the first Islamic caliphate, the Umayyad Caliphate

0	50	100		200 km
0	25	50		100 miles

Caucasus
A mountain range lying at the border of Europe and Asia. It includes seven 5000-ers, among them El'brus, Europe's tallest peak

CASPIAN SEA

Abkhazia & South Ossetia
Partially recognised republics over which Georgia has no control

Shkhara (5,201 m)
3rd highest peak in Europe and the tallest mountain in Georgia

Krubera cave
The deepest cave on Earth (2,197 m) discovered in 1960 by a Ukrainian research expedition

Joseph Stalin
Born in Gori in 1878. He is responsible for over 20 million deaths

Mud volcanoes
Azerbaijan has the highest concentration of mud volcanoes in the world

Vardzia
Medieval monastery and village built in a cave around the 12th century. Several monks still guard the holy site

GEORGIA

TBILISI

Tulip
The flower was first cultivated by Turks around AD 1000, and later introduced to the Netherlands in the 16th century

Christianity
First country to adopt Christianity as a state religion (AD 301)

AZERBAIJAN

BAKU

Irreligious Muslims
Turkey is the country with the highest irreligious Muslim population rate (73%)

Turkish Airlines
World's largest airline by the number of countries served – 106

ARMENIA

YEREVAN

Greatest chess player
Garry Kasparov was born in Baku. He remained the No.1 player from 1986 until his retirement in 2005

Neft Daslari
1st offshore oil platform

Mount Ararat △
(5,165 m)
Considered a holy mountain by Armenians

Noah's Ark
According to the Genesis flood story, the Ark landed on Mount Ararat after being afloat for 150 days

AZER.

One of the world's oldest, continuously inhabited places (founded in 782 BC). It has a scenic view of Mount Ararat

Nagorno-Karabakh
Unrecognized republic and disputed territory under the control of ethnic Armenian separatists since 1991

Lake Van
One of the largest endorheic lakes. Only one species of fish, the pearl mullet, inhabits the brakish alkaline waters

"The swimming cat"
The Van cat that lives only in the Lake Van area, is often observed to swim in the lake

Chess
Chess is a compulsory subject at schools in Armenia

Euphrates

Tigris

Caspian tigers
They were one of the largest "big cats" that ever existed. The species became extinct at the end of the 20th century – the last specimens were killed in this region

Rulers of Anatolia

Persian Empire	Greece		Roman Empire	Byzantine Empire	Ottoman Empire	Republic of Turkey
546	334	50 BC AD	330		1299	1923 Present day

ISIS
A terrorist organization, infamous for beheadings, kindnapping, pillage and brutal warfare against civilians. It gained prominence in 2014 through seizing strategic territories in Syria and Iraq. Its annual budget is around $1 billion

Kurds
Kurds are the largest ethnic minority group in Turkey (18% of the population). Until 1991, use of the Kurdish language and cultural practices were often penalized with imprisonment, while the government denied their existence, calling the nation "Mountain Turks"

Forcibly displaced people around the world

Syrian Civil War outbreak

Millions

80
60
40
20

2000 2005 2010 2015

Ancient roads
Some roads built 4,000 years ago are still in use

SYRIA

Palmyra
Ancient Semitic city with history dating back over 2,000 years. In 2015, the site and its artefacts were largely destroyed by ISIS

Euphrates

Tigris

MEDITERRANEAN SEA

"Ask not what your country can do for you. Ask what you can do for your country" – this quote was made famous by the US president John F. Kennedy but was actually by Kahlil Gibran, a Lebanese writer

Lebanon is the smallest state in continental Asia

Olive trees

The oldest living olive trees, that may be up to 2,000 years old, are found in this region

BYBLOS
First capital of the ancient civilization of Phoenicia. Continuously inhabited since 5000 BC

Baatara gorge waterfall
A cascade of 100 m high falls behind 3 levels of limestone which create natural bridges, one below the other, and then drops in to a chasm 240 m deep

Six-Day War
In 1967, this war was fought between Israel and the neighbouring countries of Egypt, Syria and Jordan. Due to a pre-emptive attack, Israel won the decisive battle. Arab casualties were over 20 times higher than those of Israel. As a result, Israel gained control of the Sinai peninsula, Gaza Strip, West Bank and the Golan Heights

BEIRUT
Nicknamed the "Phoenix", it has been ruined and built from scratch 7 times

LEBANON

In 2013, Lebanon had the world's highest population growth rate of 9.73%

Baalbek
UNESCO World Heritage site including the Temple of Bacchus, one of the best-preserved ancient temples, and six of the remaining, 19-m-high columns of the Temple of Jupiter, the biggest pagan temple in the Roman Empire

Phoenicia

This ancient civilization, which based its development on Mediterranean trade, occupied mainly the area of present-day Lebanon

Golan
Administered by Israel, claimed by Syria

Adolf Eichmann
This Nazi war criminal, who fled to Argentina after WWII, was captured by Mossad in 1960, brought to Israel and sentenced to death by hanging

Safety
Considered as one of the safest Arab countries in the world

Jews

There are between 14.7–17.4 million Jews in the world, of which 6.5 million live in Israel

Sea of Galilee
World's lowest freshwater lake (-212 m)

Milk production

Israeli cows produce on average 10,035 litres of milk per year, almost 5 times higher than the world average

Name
The name of the country comes from the river Jordan where Jesus Christ is said to have been baptized

Nobel Prize laureates

It is estimated that 22% of individuals who have received a Nobel Prize were Jewish or of Jewish descent

West Bank annexation
After the 1948 Arab–Israeli War, until 1967, the West Bank was annexed by Jordan

Hospitality
During numerous conflicts, Jordan has accepted millions of refugees from Palestine and Iraq, and recently from Syria. In 2015, there were 2.7 million registered refugees in Jordan (41.2% of the population)

Territory under Israeli-Palestinian Authority control

Pope's tree
Pope John Paul II planted an olive tree here as a symbol of peace

Bonfire
The oldest bonfire to be unearthed, dating back 300,000 years

WEST BANK

AMMAN

Hamas
Regarded by many as a terrorist organization, it has governed the Gaza Strip since 2007

JERUSALEM
Disputed capital. Destroyed twice, besieged 23 times and attacked 52 times

11% of the West Bank's population is Israeli

Mount Nebo
According to the Hebrew Bible, Moses was shown the Promised Land here

"Churchill's Sneeze"

GAZA

Dead Sea
Lowest point on earth (-429 m). It's 8 times saltier than the ocean

JORDAN

"Churchill's Sneeze"
Refers to the zigzag shaped part of the border between Jordan and Saudi Arabia. Rumour has it that Churchill, at that time the Secretary of State for the Colonies, drew the border when tipsy one Sunday afternoon in Cairo

ISRAEL

Closed borders
After Hamas gained control of Gaza in 2007, both its neighbours closed their borders. There were over 1,200 tunnels built below the border with Egypt to smuggle people, weapons and other goods

Water recycling
Most of the country is desert but thanks to advanced desalination technology, Israel is able to export fresh water to neighbouring countries and grow water-intensive produce

Israel is also the world's leader in water recycling. Around 80% of sewage is treated for land improvement

Peace with Israel
Jordan and Egypt are the only Arab countries that signed a peace treaty with Israel

Abdullah II of Jordan
King of Jordan, direct descendant of the prophet Muhammad. He holds custodianship of Islamic and Christian sacred sites in Jerusalem. This historical agreement with his dynasty dates back to 1924

Military
In 2014, Israel ranked first in the Global Militarization Index, which includes military expenditure and the number of military forces and heavy weapons

Petra, the "Rose City"
Beautiful ruins of an ancient city with buildings carved in stone, built over 2,000 years ago. It remained hidden for almost twelve centuries and was discovered in 1812

In 2007, Petra was named one of the New Seven Wonders of The World. Called "Rose City" after the colour of stone it is carved from

President Albert Einstein
In 1952, Albert Einstein was asked to become the president of Israel. He refused by saying: "All my life I have dealt with objective matters, hence I lack both the natural aptitude and the experience to deal properly with people and to exercise official functions"

Middle East
411 million people

7,207,575 km² (similar in size to Australia)

Gulf of Aqaba

"Valley of the Moon"

Wadi Rum, a desert valley cut into red sandstone and granite rock, is a popular filming location for sci-fi movies, e.g. Rogue One, The Martian and Prometheus, due to its exceptional landscape

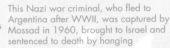

0	25	50	100 km

0	20	40	80 miles

"One Thousand and One Nights"
The collection of stories from the Islamic Golden Age, including the adventures of Ali Baba and Aladdin, were written over many centuries

Iraqi flag
The Arabic inscription on the Iraqi flag reads "Allahu Akbar" meaning "God is great"

CASPIAN SEA

Islam
Islam is currently the 2nd largest religion by population and the fastest growing. It is expected to surpass Christianity by the end of the century

Two nations
Iraq's two largest ethnic groups are the Arabs and Kurds. The latter live mostly in Iraqi Kurdistan, the country's only autonomous region

Yazidi bombings
The 2nd most deadly terrorist attack. 4 bomb attacks killed 796 people in 2007

Halabja chemical attack
In 1988, Saddam Hussein ordered a chemical attack on the Kurds, killing up to 5,000. The US condemned the act while still supporting Hussein

Mesopotamia
The fertile Tigris–Euphrates river system is the birthplace of the world's first civilization. Mesopotamian inventions include writing, irrigation, the wheel, sailboats, the concept of counting, written set of laws and astrology

Ancient Babylon
Believed to be the 1st city in history to reach a population of 200,000

BAGHDAD

Tower of Babel
A mythical structure where different languages were meant to originate, associated by some historians with the Ziggurat of Babylon

IRAQ

Tigris

Euphrates

Oil
Historically, Saudi Arabia has been the largest oil exporter and producer with the 2nd largest proven reserves

Wadi Al-Salaam
With over 5 million bodies, it is the largest cemetery in the world

Desalination
Kuwait was the first country to introduce desalination as a large-scale solution to their water supply problem

Low mortality
UAE and Qatar have two of the lowest mortality rates in the world, 1.53 and 1.99 per 1,000 population respectively

KUWAIT
KUWAIT

"Oil population"
Since the discovery of oil in the 1930s, the population has increased more than 10 times

No rivers
The largest country in the world to have no permanent rivers

Bahrain's population
Around half of Bahrain's 1,377,000 population is foreign-born

The Gulf
World's warmest sea. Holds 60% of the world's oil reserves

Wealthiest
Qataris have the highest income per capita (at purchasing power parity), surpassing the 2nd richest nation by around 30%

Sex ratio in UAE
Of all countries, Emirati women are the most outnumbered by men: 2.19 men per 1 woman

0% of the ish species from the Red Sea can be found nowhere else on Earth

Jeddah Tower
It will be the first man-made structure in history to reach 1 km in height. Completion is planned for 2020

Arabian Peninsula
The largest peninsula in the world; similar in size to India

Garden of Eden
Bahrain is believed to be this biblical place

BAHRAIN
MANAMA

OMAN

QATAR
DOHA

Gulf of Oman

ABU DHABI

MUSCAT

World's northernmost tropical sea with one of the highest surface water temperatures, around 34°C

MEDINA
The 2nd holiest city in Islam, it was the first capital of the Muslim Empire

RIYADH
The capital's camel market is one of the largest in the world and sells around 100 camels per day

UNITED ARAB EMIRATES

Horse breeders
Omanis are known for breeding the finest Arab horses

Hegira
The Great Emigration of the Prophet Muhammad and his followers from Mecca to Medina

SAUDI ARABIA

Muhammad
The most common given name in the world

محمّد

RED SEA

MECCA
The birthplace of Islam and home to the holiest shrines in the Islamic world

Forbidden City
Entry to the city of Mecca is strictly forbidden for non-Muslims

Empty Quarter
The Rub' al Khali is the largest contiguous sand desert in the world. Its dunes are up to 250 m in height, more than twice the size of the Great Pyramid of Giza

OMAN
The only member of the Gulf Cooperation Council to oppose Saudi intervention in the Yemeni Civil War, Oman helped displaced Yemenis by sending in supplies and taking in refugees

The sea is still widening due to tectonic plate movement

Women's right to vote
In the 2015 elections, women were allowed to vote and run for office for the first time. They are still not permitted to drive

Lone republic
The only republic in the Arab peninsula. Other countries are kingdoms or emirates ruled by one family or one ruler

Shipbuilders
Omanis were once considered the greatest shipbuilders – they used "coir", rope made from coconut fibre

Just 6 cm of rainfall per year and no major source of fresh water result in evaporation of up to 2 m per year

Fertile lands
The mountains and highlands attract rainfall, making Yemen the most fertile area of the Arabian Peninsula

YEMEN
The poorest country in the Middle East and ranks among the top 5 most corrupt

SANAA

Marriage
Men in Yemen, Saudi Arabia, India and elsewhere still insist that it is their Islamic right to marry girls younger than age 15. Yemeni Muslim activists argue that some girls are ready for marriage at age 9

SOCOTRA (Yemen)

Gulf of Aden

ARABIAN SEA

100   200   400 km

75   150   300 miles

Leaders of the Soviet Union

Vladimir Lenin	Joseph Stalin	Georgy Malenkov	Nikita Khrushchev	Leonid Brezhnev	Yuri Andropov	Konstantin Chernenko	Mikhail Gorbachev
1922 1924		1953 1955		1964		1982 1984 1985	1991

Horses
First domesticated in the western part of the Eurasian Steppe between 4000–3500 BC. Today, horsemeat is very important in Kazakh cuisine

Transcontinental country
Technically, Kazakhstan is a transcontinental country. Its eastern part lies in the European part of Eurasia

Boxing
Kazakhstan is well known for its boxers who have won 7 gold medals at the Olympics

Corruption
In 2005, the World Bank listed Kazakhstan as a corruption hotspot, together with Angola, Bolivia, Kenya, Libya and Pakistan

9/11
After the 2001 terrorist attack, Kazakhstan offered military bases and air space to help fight al-Qaeda

Herding
Before incorporation into the Soviet Union, animal herding was the mainstay of the Kazakh nomadic population. Since then, the overuse of chemical fertilizers and groundwater resources has depleted the soils significantly

KAZA

Baikonur Cosmodrome
World's first and largest operational space launch facility. Leased to the Russian Federal Space Agency until 2050. World's first orbital spaceflight and first manned spacecraft were launched from here

Aral Sea
After the rivers feeding the Aral Sea were diverted by Soviet irrigation projects in the 1960s, the lake has declined to around 10% of its original size

Caspian Sea
The largest lake in the world, almost 5 times larger than the 2nd largest, Lake Superior

Aral Sea tributaries
During the Soviet era, Syr Darya and Amu Darya were extensively used for farm irrigation which almost led to the disappearance of the Aral Sea

Syr Darya

Ethnic groups
More than 120 different ethnic groups live in Kazakhstan, deported during Stalin's rule

CASPIAN SEA

MO'YNOQ
Once a busy port city, its remnants now lie in a desert wasteland

Basmachi Revolt
In 1916, Muslim people of Central Asia revolted against the Russian Empire when it started drafting Muslims for World War I. The rebellion was crushed

UZBEKISTAN

Muruntau gold deposit
The largest gold deposit in Eurasia

Amu Darya

Garabogazköl Aýlagy
World's largest deposit of natural marine salt. Rusty machinery on movable railway tracks scrape the salt off the lagoon's surface

Water source
Turkmenistan has only one source of fresh water – the Amu Darya river

Classification
The Caspian Sea is considered a lake because it has no outflows. It became landlocked around five million years ago

"The Door To Hell"
A gas field set on fire by Soviets has been burning for over 40 years

Doubly landlocked
One of only two doubly landlocked countries in the world (surrounded by landlocked countries only) along with Liechtenstein

Oil
When oil exploration started here in 1873, it was one of the largest known fields of this resource

Free resources
Water and gas is free in Turkmenistan since the country has the world's 4th largest natural gas reserves

TURKMENISTAN

Amu Darya

● ASHGABAT

President Niyazov
One of the world's most totalitarian dictators imposed personal eccentricities upon the country, such as renaming months

Karakum Desert
70% of the land is covered by this Black Sand desert

0	50	100	200		400 km

0	25	50	100		200 miles

Uranium
Kazakhston produces almost 40% of the world's uranium, making it the largest producer

Apple
The original wild ancestor of the apple was found growing in this area

Traditional production
Kazakhstan has traditionally specialized in wool and grain production

Russian settlers
During the 1890s, Russian settlers expropriated the fertile lands of Kazakhstan and displaced the nomads living there. Starting in 1906, around half a million Russian farms were set up

Irtysh

• ASTANA
Since its foundation in 1830, the city has changed its name several times between Akmoly, Akmalinsk, Tselinograd, Akmola and Astana

IMF debt
Kazakhstan became the first former Soviet republic to repay all of its debt to the International Monetary Fund in 2000, 7 years before it was due

Irtysh

Saiga antelope
Around 40% of these endangered animals suddenly died in Kazakhstan in the course of a few weeks. Pasteurellosis, a respiratory disease, is suspected to be the cause

Natural resources
The country is rich in coal, oil, gas and rare metals

Kazakh and Tajik women
Both gained the right to vote in 1924

ISTAN

Size
The largest landlocked country in the world, 9th largest country by size, a huge place with a small population of only 17.6 million people

Lake Balkhash
The eastern part is deeper and much saltier than the western, fresh water part. The narrow strip of land between them, evaporation differences and the river Ili are reasons for this anomaly

Nuclear arsenal
When the Soviet Union collapsed, Kazakhstan returned warheads to Russia and destroyed its nuclear testing infrastructure

Kyrgyzstan's flag
The sun on the flag symbolizes prosperity and its 40 rays represent the number of tribes united by the national hero, Manas, in a fight against the Mongols

Snow leopard
They live in the mountains in the east of the country and are considered a national symbol. Kazakhs revere the animal for its bravery, independence and intelligence

Radiation
High radiation after the Soviet Union's nuclear testing and some 30 uranium mines have caused radioactive fallout. Kazakhstan has tried to persuade China to stop their atomic testing near its territory

• BISHKEK
The city's name was changed from Frunze in 1991, after independence

Ysyk-Köl
Meaning "Warm lake", it is the 2nd largest saline lake in the world. Although surrounded by snow-capped peaks, it never freezes

Syr Darya

Arslanbob
Wild walnut forest that produces 1,500 tonnes of walnuts annually, making it Kyrgyzstan's major natural resource and a national treasure

• TASHKENT

KYRGYZSTAN

Partition of the Soviet Union in 1991

Largest post-Soviet countries

	Area (km²)	Population
Russia	17,075,400	143,457,000
Kazakhstan	2,717,300	17,625,000
Ukraine	603,700	44,824,000

High elevation
At 3,186 m, Tajikistan has the 3rd highest average elevation above sea level

TAJIKISTAN

• DUSHANBE

Marco Polo sheep
They have the longest, spiralling horns of all sheep. The longest ever recorded was 1.9 m and weighed 27 kg

Iskanderkul
Mountain lake named after Alexander the Great – "Iskander" for Alexander and "kul" for lake

Poverty
After gaining independence, the civil war and recurring natural disasters resulted in Tajikistan becoming one of the poorest countries. In some regions, over 70% of the population live in poverty

Estonia
Latvia
Lithuania
Belarus
Ukraine
Moldova
Georgia
Armenia
Azerbaijan

Russia

Kazakhstan
Kyrgyzstan
Tajikistan
Turkmenistan   Uzbekistan

Beluga (sturgeon)
World's largest freshwater fish, critically endangered. It lives up to 118 years

Caviar
Beluga caviar is the most expensive in the world, reaching $10,000 per kilo. Until the mid-2000s embargo, Iran was the largest producer of the valuable roe

CASPIAN SEA

Nuclear fuel cycle

Mining
→
Conversion
→
Reactor ← Enrichment → Uranium bomb
Reproscessing
Reprocessing → Plutonium bomb

Lake Urmia
Until the 1980s, it used to be the 6th largest salt lake in the world. In just 2 decades it decreased in size by 90% due to climate change and damming projects

Mount Damavand (5,601 m)
The highest mountain in Iran and Asia's highest volcano

Operation Ajax
Coup d'état, staged by British Intelligence and the CIA in 1953, against democratically elected prime minister Mohammad Mosaddegh, who wanted to limit foreign involvement in oil production in Iran

Gas & oil pipelines

Elburz Mountains

TEHRAN
During one of his travels, around AD 1270, Marco Polo claimed to have seen the tombs of the biblical Three Kings in this region

"Argo"
From 1979-81, 52 American diplomats and citizens were held hostage for 444 days in the Tehran US Embassy. The movie "Argo" about this incident won the Academy Award for Best Picture

Irangate scandal
During the Iran-Iraq War in the 1980s, the US officially supported Iraq, but secretly sold weapons to Iran to finance Contra rebels in Nicaragua. Iran was at the time subject to an arms embargo that the US had imposed themselves

Gas & oil pipelines

Arak ★
Heavy water reactor and production plant

Fordow
Uranium enrichment plant

Nuclear safety
Iran is currently the only nuclear country that has not signed the Convention on Nuclear Safety

Afghan refugees
Due to the Soviet War in Afghanistan, around 6 million Pashtuns were forced out of their homes. Half of them found refuge in Iran. It's the 4th largest displaced population

★ **Natanz**
Uranium enrichment plant

★ **Isfahan**
Uranium conversion plant

Iranian Revolution
In 1979, Mohammad Reza Shah, the Iranian monarch supported by the US, was overthrown, the Islamic Republic of Iran was installed and Ayatollah Khomeini became its supreme leader

Iran's oil reserves
The country ranks 4th with roughly 10% of the oil on the planet

IRAN

Hand-woven carpets
Iran's 2nd export product after oil. The world's largest handmade carpet produced here was the size of a football field and was made for the Abu Dhabi mosque

Bushehr
Nuclear power station

Sex change operations
Although homosexuality is punishable by death under Sharia Law, surgery for changing one's sex is legal here

Iran is 2nd only to Thailand in the number of sex change operations conducted per annum. Half the cost is funded by the state

Organs trade
The only country where selling one's kidney is legal. Around 1,400 Iranians sell their kidneys every year and the average price ranges from $2,000 – $4,000

The Gulf
Its average depth is just 35 m. It formed just after the last glaciation. Until 15,000 BC it was entirely dry

Deadliest blizzard
The blizzard that hit Iran in 1972 was the deadliest in history as a week of low temperatures and snow storms resulted in the deaths of 4,000 people

Human trafficking
Together with Saudi Arabia, North Korea and Dem. Rep. Congo, Iran is considered a hub of human trafficking

Natural gas
Iran holds around 15% of the world's proven natural gas reserves. It's 2nd only to Russia which owns over 1/4 of the planet's resource

Rape
Rape is a crime in Iran, however, in order to prove a rape, a victim has to present several male witnesses

First Persian Empire
One of the largest empires in history, founded by Cyrus the Great in 550 BC and conquered in 330 BC by Alexander the Great

Iran's major gas fields

Gachin ★
Uranium mine

Female football
The Iranian national women's soccer team could not take part in the 2012 Olympics qualifier game after FIFA banned the hijab

South Pars/North Dome field
The largest natural gas field containing around 19% of the Earth's proven natural gas. It is shared between Iran and Qatar

| 0 | 75 | 150 | 300 km |
| 0 | 50 | 100 | 200 miles |

Opium production area in Afghanistan

km²

3,000

2,000

1,000

0

1994 1996 1998 2000 2002 2004 2006 2008 2010 2012 2014 2016

Year

Soviet-Afghan War
The only war ever lost by the Soviet Union (1979–89). It left up to 2 million people dead and many millions still displaced today

Operation Cyclone
The CIA armed, facilitated training and funded Jihadi warriors, the Mujahideen, with over $20 billion to fight the Soviets. The Mujahideen included Osama bin Laden and many others who later joined groups such as al-Qaeda and the Taliban

K2 (8,611 m)
World's 2nd highest mountain with the 2nd highest fatality rate (25% of attempts result in death). Only around 300 people have reached the summit. The mountain has never been climbed during winter

HIMALAYA

Broad Peak (8,047 m)
First winter ascent in 2013 by 4 Polish climbers. Two of them died when descending

BALKH
Ancient centre of Buddhism until the Arab invasion in the 7th century

Silk Road
Afghanistan was an important part of the Silk Road and meeting point of the greatest civilizations

Gilgit-Baltistan
Administered by Pakistan, claimed by India

Indus

Buddhas of Bamiyan
Over 1,500-year-old Buddha statues destroyed in 2001 by the Taliban. Buddhist paintings in the Bamiyan caves are the oldest known surviving examples of oil painting

Nanga Parbat (8,126 m)

Missing limbs
Around 60,000 people are missing limbs – most of them were injured by landmines

KABUL
Zablon Simintov, an Afghan carpet trader, is the keeper of the last synagogue in Kabul and one of the last Jews living in Afghanistan

Gasherbrum II (8,035 m)
Gasherbrum I (8,068 m)

Indus

Bin Laden's death
The terrorist was killed here in 2011. There was a $25 million bounty placed on him by the FBI. He was buried at sea in an undisclosed location

AFGHANISTAN

American War in Afghanistan
The longest war waged by the US in the country's history (2001–14)

Football production
Most of the world's footballs are made in Sialkot, often with the use of child labour

ISLAMABAD

Khewra Salt Mine
The 2nd largest salt mine in the world, which produces pink, so-called "Himalayan salt". It was discovered around 320 BC, during the conquest of Alexander the Great

"The Afghan Girl"
The 1985 National Geographic magazine cover photo became probably its most iconic

No railway
There is no passenger rail transport in Afghanistan

Helmand Province
Around half of Afghanistan's opium is produced here

Opium
Afghanistan is the world's largest opium producer. Around 20% of the country's economy is driven by its production

Indus river
It's one of the longest rivers in Asia, 3,180 km long, and a vital element of Pakistan's economy. Its source is located in Tibet

Malala Yousafzai
At the age of 17, she became the youngest Nobel Peace Prize laureate for advocating female education. She is a survivor of a murder attempt by the Taliban

Partition of India
In 1947, India and Pakistan split up. It resulted in the forced migration of 14 million Muslims and Hindus. Around 1 million died of exhaustion

Irrigation
The Indus supports the largest, continuous agricultural irrigation system on the planet spanning over 144,000 km² along the river

Indus river dolphin
It was the first side-swimming cetacean to be discovered. This threatened species of sweet water dolphin inhabits waters of the longest river in the region

Indus

Edhi Foundation
It is an NGO in Pakistan which provides the largest private ambulance network in the world

PAKISTAN

Booming population
Shortly after Pakistan's inception, in 1950 it had a population of a mid-sized European country, 37.5 million. Today, it's the 6th most populous country in the world with 189 million citizens

Harappa cotton
The first people in Eurasia to cultivate cotton were the Harappa civilization from around 3300 BC. They vanished around 1600 BC and left no written records

Thar Desert
The most densely populated desert in the world (83 people/km²). In contrast, the Sahara has one of the lowest population densities (1 person/km²)

Languages
The national language of Pakistan, Urdu, is spoken as a first language by just 7.5% of the population

English is the official language used by the government, military, social services and most schools. After India, Pakistan is the 2nd largest English-speaking country outside the United States

Gwadar Port
One of the largest deep water ports in the world, and an important point of the Chinese Maritime Silk-Road project

Tharparkar
Known as the only fertile desert in the world. Two perennial streams feed it during the rainy season

Kashmir
Beautiful land of green mountain valleys and saffron flowers. It remains a conflict zone between India, China, Pakistan and the people of Kashmir

Jammu and Kashmir
Administered by India, claimed by Pakistan

Indus Valley Civilization
Established in the Bronze Age, between 3300 and 1300 BC. It created the first urban centres in the region, including the development of the world's first urban sanitation system

It was the most widespread of three civilizations of the "Old World" (the others being Egypt and Mesopotamia)

Growing population
India's population is expected to surpass China's and become the largest on the planet with over 1.4 billion citizens by 2022

Economic growth
India's economic policies, initiated in 1991, have turned its economy towards services and industry and opened the market to private and foreign investors. In 2016, India's economy was 7th largest in the world

DHARAMSALA
Home to the Dalai Lama and the Tibetan government-in-exile

Kumbh Mela Festival
One of the largest peaceful gatherings in the world. 120 million people spent 2 months on a pilgrimage in 2013 to bathe in sacred rivers

Lotus Temple
Famous place of prayer open to all religions

NEW DELHI
The city was founded in 1911. Before then, Calcutta (Kolkata) was the capital

NEPAL

Karni Mata Temple
Home to over 20,000 rats, worshipped as reincarnated family members of a deity

JAIPUR
The city was painted pink to welcome the Prince of Wales, later King Edward VII, in 1876

Taj Mahal
Built between 1632 and 1648 by emperor Shah Jahan to house the tomb of his favourite wife

Dhaulagiri I (8,167 m)

Economic growth
India has the 2nd fastest growing economy after China, 6.1% per year

Company rule in India
For 101 years, beginning in 1757, large parts of present-day India were ruled by the British East India Company

Gandhi's Salt March
Symbolic milestone on India's path to independence. It was an act of mass civil disobedience, manifested through non-violent protests against the British salt monopoly

Tiger
There are only around 3,890 tigers alive in the wild today, down from around 100,000 in 1900. 70% live in India

Kanha Tiger Reserve
The most famous of India's 50 tiger reserves

Hinduism
One of the oldest religions. Around 1 billion people are Hindu

INDIA

Bollywood
India has the biggest film industry in the world by annual admissions. In 2016, more than 200 Hindi films opened in Indian cinemas. The industry is based in Mumbai

ARABIAN SEA

Largest democracy
With 1.3 million people, India is the largest democracy in the world

Food production
India is the world's largest producer of bananas, lemons, mangoes and spices, such as ginger, among others

MUMBAI
With 21.7 million people living in the urban agglomeration, it is the 2nd most populous city in India after Delhi

Chilkur Balaji Temple
There is a belief that the Balaji Temple's deity has the power to grant US visas. Many come here for his blessings before their visa interviews

Dairy
India is the largest producer of milk and butter

Ancient civilization
The only island group in the Central Indian Ocean region where an ancient civilization flourished

The smallest Asian country by population and area

Goa
The smallest state in India with the highest per capita income. It is famous for its beautiful beaches and rich flora and fauna

Kollur Mine
The 1st and up until the 1890s the only diamond mining centre in the world

MALDIVES

Lowest average elevation (1.5 m) of all countries

MALE
With an area of just 5.8 km², it is one of the most densely populated cities in the world (26,535 people/km²)

Doctors & engineers
India has the highest number of graduate doctors and engineers

Lowest highest point
World's lowest highest point, 2.4 m above sea level

Kerala rubber
India is one of the world's largest natural rubber producers and around 90% comes from the state of Kerala

Deccan peninsula
World's 2nd largest peninsula

Cinnamon
Sri Lanka is the largest producer of cinnamon with around 70% of global production

0   10 km
0   10 miles

INDIAN OCEAN

Kerala backwaters
Over 900-km-long chain of lagoons and lakes, home to many unique aquatic species

SRI LANKA

0   125   250   500 km
0   75   150   300 miles

SRI JAYEWARDENEPURA KOTTE

Buddha's birthplace
Buddha was born in 623 BC in the holy area of Lumbini

Annapurna I (8,091 m)
The first eight-thousander to be climbed and the most dangerous with the highest fatality rate, 32%

Manaslu (8,163 m)

Kangchenjunga (8,586 m)

KATHMANDU
The largest city in the Himalaya (1 million people)

Ganges

Ganges river
Third largest river by discharge and one of the most polluted in the world. The most sacred river for Hindus

BANGLADESH

DHAKA

KOLKATA
The Missionaries of Charity was established in Kolkata by Mother Teresa in 1950 to help the sick and needy

Green country
Bhutan is considered a model for environmental conservation policies. At least 60% of its land must stay under forest cover

"Gross National Happiness"
Bhutan established this non-materialistic measure of a country's growth

Tourism
Foreigners were not allowed to enter Bhutan until 1974

BHUTAN
THIMPHU

Darjeeling Himalayan Railway
Famous narrow-gauge railway opened in 1881, developed by the British for a health resort and sanatorium

CHERRAPUNJI
Credited as the place with the highest recorded rainfall in a year (26,461 mm)

Tree bridges
Rubber trees in Cherrapunji have grown in such a way as to form bridges. Some bridges are over 500 years old

Muhammad Yunus
This Bangladeshi entrepreneur and professor received a Nobel Peace Prize for developing the concept of "microcredit" and creating the Grameen Bank which grants small loans to the poorest and helps them break out of poverty

Population density
Bangladesh has the highest population density in the world (1,118 people per km²), not counting "small" countries (less than 2,000 km² in area)

Agriculture
66% of Bangladeshi land is cultivated, the highest in the world

1970 Bhola cyclone
The deadliest tropical cyclone ever recorded, with up to 500,000 casualties, mostly in Bangladesh, and one of the deadliest natural disasters in modern history

Influence of the Himalaya on the climate

Altitude in m / Himalaya / Rain shadow / Monsoon / Gobi / Distance in km

Influence of the Himalaya on the climate
The tallest mountain range on Earth forms a natural barrier to the monsoon, creating a wet, subtropical climate over India's land mass and a dry climate for the Tibetan plateau and Gobi desert

Monsoon
In the summer, the temperature on the land rises much faster than that of the ocean, creating a significant seasonal difference in air pressure. As a result, the moist air above the ocean is sucked inland, to later cool down and condense as rain as it rises towards the Himalaya. The monsoon creates around 90% of precipitation in the region

Journey of the Indian subcontinent
At the time the dinosaurs became extinct, India was an island travelling across the ocean, several centimetres each year, towards Eurasia

Colossal collision
The journey had taken around 90 million years. When it collided with Asia's land mass, around 10 million years ago, the youngest and the tallest terrestrial mountains on Earth were formed, the Himalaya

BAY OF BENGAL
The largest bay in the world

Penal colony
From the late 18th to early 20th centuries, the Andaman Islands were used as a British penal colony

ANDAMAN ISLANDS (India)

Coastal migration
There is a theory that the Andaman Islands were a key stepping stone in early human migration from Africa to Australia and Indonesia

Sentinelese
Indigenous people who purposely resist contact with the outside world

Suicide
Sri Lanka has the highest suicide rate – 34.6 per 100,000 people

Flag
The Sri Lankan flag is the only one to recognize different religious groups. Different colours represent Buddhist, Muslim and Hindu Tamil communities

Police massacre
In 1990, an estimated 774 unarmed Sri Lankan Police officers were killed by a separatist organization, the Liberation Tigers of Tamil Eelam

Pinnawala Elephant Orphanage
The largest herd of captive elephants

Plate collision
The island arc of the Andaman and Nicobar Islands was created when the Indo-Australian Plate and Eurasia collided, the same collision that formed the Himalaya

ANDAMAN SEA

NICOBAR ISLANDS (India)

Malaria
During the 17th and 18th centuries, the Nicobar Islands were abandoned a couple of times due to outbreaks of malaria

Asian dust
Seasonal meteorological phenomenon. Strong winds blow up clouds of dry soil from Mongolia, China and Kazakhstan and carry it over North Korea, South Korea, Japan and Russia

Mongol Empire
The largest contiguous empire in history (1206–1368). At its greatest extent it had an area of 24,000,000 km², much larger than present-day Russia

Camel
The two-humped Bactrian camel comes from Mongolia and surviv temperatures from -40°C to 40°C

Borders
China borders 14 countries (not counting Hong Kong and Macao), more than any other state, apart from Russia

Living in caves
It is estimated that around 30 million Chinese live in caves – warm in the winter and cool in the summer

Agriculture and industry
China has the highest agriculture and industry sector output (as a percentage of GDP) of all countries

Eating snow
Camels living in the Gobi desert eat snow, because often it's the only source of water. Snow evaporates quickly as the sun rises

"Empty lands"
Lowest population densit of all sovereign states – 1.92 people per km²

Food production
China is the largest producer of potatoes, tomatoes, wheat, garlic, peaches, watermelons, walnuts, carrots, onions, lettuce, grapes, honey, mushrooms and many other foods

Motor vehicles
28,118,794 motor vehicles were produced in China in 2016, more than anywhere else in the world

Gobi desert
Rain shadow desert formed as the Himalaya range blocks rain-carrying clouds from the Indian Ocean

MON

CHINA

K2 (8,611 m)
Broad Peak (8,047 m)
Gasherbrum II (8,035 m) —
Gasherbrum I (8,068 m)

Mobile phones
There are 1,321,930,000 mobile phones in China

Renewable energy
China in a leader in wind turbine and solar panel production. It is also the largest renewable energy producer

Aksai Chin
Administered by China, claimed by India

Taiping Rebellion
One of the bloodiest conflicts in history (1850–64). Up to 30 million people died during this civil war fought by the Taiping Heavenly Kingdom, an opposition state trying to overthrow the Qing dynasty

Exports
China's export value is $2,011,000,000,000, more than any other country. Only the EU has a higher value

Huang He
(Yellow)

Chang Jiang
(Yangtze)

Jerzy Kukuczka
He was a Polish mountaineer, the 2nd man to climb all 14 eight-thousanders (after Reinhold Messner) and the person who established 9 new routes on eight-thousanders, more than anyone else

Himalaya
Highest mountain range on Earth. Includes nine 8,000 m peaks and fifty higher than 7,200 m

No oxygen
Above the level of 8,000 m there is around 66% less oxygen than at sea level

CO_2

Huanglon
UNESCO World Heritag site, known for its colou terraced water poo formed by calcite depos

Himalaya

Cho Oyu (8,201 m)
Considered the easiest eight-thousander to climb to the summit

Xixabangma Feng (8,027 m)
Lhotse (8,516 m)
Makalu (8,463 m)

Mekong

Wolong Nationa Nature Reserve
Hosts numerous endangered species including the largest population of giant pandas

Mount Everest (8,848 m)
The highest mountain on Earth. The first ascent was by Edmund Hillary and Tenzing Norgay in 1953. The first winter ascent was by Leszek Cichy and Krzysztof Wielicki in 1980. It was also the first winter ascent of any eight-thousander

1959 Tibetan uprising
Brutally crushed by the forces of the Communist Party of China. Around 80,000 people were killed and most monasteries were destroyed

Eight-thousanders
There are 14 peaks on Earth higher than 8,000 m. The first person to climb them all was the Italian Reinhold Messner, the last climb achieved in 1986

Tea
Tea plants probably originated in the region of north Myanmar and southwest China

Tibet Autonomous Region

0   150   300   600 km

0   100   200   400 miles

Przewalski's horse
These Mongolian native horses have two chromosomes more than an average horse. They are the last truly wild horses left on the planet

Population
1,383,925,000 people live in China, more than in Africa (1.2 billion) which is the 2nd most populous continent after Asia

Urbanization
Over 50% of people live in urban areas (an increase from 20% in 1990). There are over 160 cities in China with a population greater than 1 million

There are as many Chinese today as there were people on Earth around the year 1875

ULAN BATOR
Founded in 1639 as a nomadic Buddhist monastic centre. It changed its location 28 times, with each location being chosen ceremonially. It settled here in 1778

Population boost
Until the 1960s, the government encouraged families to procreate. Mao Zedong believed that the country's power was in population growth. It grew from around 540 million in 1949 to 940 million in 1976. In this period, life expectancy also increased from around 35 to 66 years

One-child policy
Introduced in 1979, inspired by a global overpopulation debate and rejected in 2015. The policy is thought to have prevented 400 million births

Dinosaur eggs
Roy Chapman Andrews, the alleged inspiration for Indiana Jones, made the first discovery of dinosaur eggs at the Flaming Cliffs site

Panjin Red Beach
One of the world's largest wetlands. In autumn, the seepweeds growing there take on a crimson hue, resulting in a sea of red

Population growth in China

Largest companies
52 out of the world's 500 largest companies have headquarters in Beijing, more than any other city in the world

Yellow river
In its basin, ancient Chinese civilization was born. Called the "River of Sorrow" after the flood of 1931 which killed up to 4 million people

BEIJING
It will be the first city to host both Summer (2008) and Winter (2022) Olympics

BO HAI

Rice
China is the largest producer of rice (204,300,000 tons annually) and its largest consumer

Huang He (Yellow)

Intentional flood
In 1938, the Chinese army breached a dyke on the Yellow river to stop the Japanese military offensive. It took up to 900,000 lives and destroyed thousands of villages

YELLOW SEA
Sand from the Gobi desert gives it its characteristic yellow colour

Terracotta Army
Over 8,000 detailed, life-size sculptures of warriors created for and buried with the emperor Qin Shihuangdi as protection in his afterlife. The army lay underground for over 2,000 years

Spring
Temple Buddha
The tallest statue in the world (128 m). It was built between 1997 and 2008

Danyang–Kunshan Grand Bridge
This 164.8-km-long high-speed railway bridge is the longest bridge in the world

Yangtze river
The longest river in Asia (6,380 km). Its river basin is home to around 1/3 of China's population

Chang Jiang (Yangtze)

SHANGHAI

World's busiest port
Shanghai port is the world's busiest container port by container traffic. In 2010, it overtook the port of Singapore

Three
Gorges Dam
World's largest power plant by capacity (22,500 MW). Its construction caused many ecological changes, flooded historical sites and forced the displacement of around 1.3 million people

DATANG
Known as "Sock City". Up to 8 billion pairs of socks are produced yearly here, around 1/3 of global production

EAST CHINA SEA

"The Great Leap Forward"
Economic reform (1958–62) that rapidly industrialized the country, at the expense of agriculture. It led to the 1959 Great Chinese Famine that killed an estimated 15–36 million people

Paper
The papermaking process developed in China around the 2nd century AD. The original formula has been lost, but the process included use of bark, hemp, silk and fishing net

TAIWAN

TAIPEI

High-tech capital
Taipei is a dynamic economic centre, with its growth based on high-tech industry and exports. The city's Taipei 101 Tower was the highest building on Earth when it opened in 2004

Gambling
Macao is one of the world's largest gambling markets and the only place in China where casinos are legal

Transfer of sovereignty
Hong Kong and Macao were transferred back to China in 1997 and 1999 respectively

Yu Shan (3,950 m)
This mountain makes Taiwan the 4th highest island in the world

Taiwan Strait

PACIFIC OCEAN

Autonomous territory and the world's most densely populated region (18,534 people per km²)

MACAO

HONG KONG
Autonomous territory with the highest number of skyscrapers in the world – around 8,000 buildings with more than 14 floors

China claims Taiwan as its 23rd province

Outside the UN
Taiwan is the most populous state and the largest economy of all the states which are not members of the United Nations

Population: 23.5 million people ≈ Australia
GDP: $528.5 billion ≈ Argentina

HAINAN DAO

SOUTH CHINA SEA

History of the conflict

1910 Korea is annexed and occupied by the Empire of Japan

1945 After the Japanese defeat in WWII, Korea is divided into two zones occupied by the US and the Soviet Union

1948 The zones fail to reunite and two separate governments are appointed

1950 After the North Korean invasion of South Korea, the Korean War (1950–3) breaks out. The north is supported by China and the Soviet Union while the South is backed by the US and United Nations

Juche
North Korea's communist ideology that incorporates concepts of Marxism–Leninism

Religion in North Korea
Although the country claims to be a secular state, hundreds of thousands have been persecuted for their religious beliefs. All religious groups are in fact in conflict with Juche ideology

Dictatorship
North Korea has the lowest level of press freedom, the lowest level of democracy and is the most corrupt country in the world according to the Corruption Perceptions Index

Kimchi
Traditional Korean side dish, made of fermented vegetables. Koreans consume around 18 kg of Kimchi annually per capita

Controversial dog meat
The history of dog meat in the Korean diet goes back about 2,000 years. Although only a small part of the population eats it regularly, it is still available in some Korean restaurants

Military
North Korea has the largest military budget as a percentage of the country's GDP (22.9%) and the highest ratio of active troops per capita (4.8%)

Kim Jong-un
The North Korean dictator is currently the world's youngest president (33 years old)

North Korean calendar
Instead of following the Gregorian calendar, North Korea counts the years of a new era from dictator Kim Il-sung's date of birth – 15 April 1912, the same day the Titanic sank

Just appearance
Construction of the 105-storey Ryugyong Hotel started in 1987. After 30 years, it remains unfinished and empty

NORTH KOREA
Since the mid-20th century, it has been the world's most politically isolated country

Pueblo crisis
North Korea is the only country which holds a US Navy ship captive. It was attacked here, in 1968. The crew were imprisoned and tortured for 11 months before being released

PACIFIC OCEAN

Rungrado 1st of May Stadium
The largest stadium in the world with a total capacity of 114,000 seats. Famous for huge performances celebrating the dictator

Mansudae Art Studio
Employs nearly 4,000 workers, responsible for creating the most notable export product of North Korea – giant sculptures

PYONGYANG

Demilitarized Zone
Contrary to its name, it is one of the most militarized places on Earth. The Korean peninsula was divided into North Korea and South Korea in 1948, approximately along the 38th parallel

Kijong-dong
Officially, a village housing about 200 families – in fact, an uninhabited propaganda town featuring anti-West loudspeaker broadcasts

Hangul
한글
The alphabet is one of the last things that Koreans have in common. It was created by King Sejong the Great in 1443

Flag pole competition
North Korea and South Korea are competing for the highest flag pole. They are currently 160 m and 100 m high respectively

Songdo International Business District
Built from scratch for over $40 billion, it's the world's first complete smart city

SEOUL
The first city in East Asia to have electricity – first introduced to the royal palace in the 1880s by the Edison Illuminating Company

Crime re-enactment
Violent convicts in South Korea may sometimes be taken to the crime scene, handcuffed and forced to recreate the criminal act. Date and place of the re-enactment is made public and the ritual is documented by the media

Taekwondo
Korean martial art, characterized by sophisticated kicking techniques. South Korea is the most successful country in this sport. It's the only Olympic sport which originated in Korea

SOUTH KOREA

Innovation
South Korea has the most innovative economy according to the Bloomberg Innovation Index

Seoul - Jeju
The busiest air route in the world, used by 11 million passengers in 2015

Seoul - Jeju (450 km)

Cheomseongdae
Probably the oldest surviving space observatory on the planet, built over 1,300 years ago, and made of 362 granite stones, corresponding to the 362 days of the lunar calendar

Shipbuilders
41% of the world's ships are build in South Korea. Four of the world's largest shipbuilding companies are from here: Hyundai Heavy Industries, Daewoo Shipbuilding, Samsung Heavy Industries and Hyundai Samho Heavy Industries

YELLOW SEA

Fastest internet
The country has the highest average internet connection speed

Korea Strait

Internet adoption and addiction
At the beginning of 2017, nearly all households in South Korea had internet access

Almost 100% of South Korean youths have a smartphone. They spend about 20 hours per week playing video and mobile games

It's estimated that over 0.6 million children in South Korea aged 10-18 are addicted to the internet and games

Female divers
The well-being of families on Jeju once relied strongly on Haenyeo, female divers, whose tradition has passed down the generations since the 5th century, but has declined since industrialization in the 1960s

Hawaii of the Koreas
Recently, Jeju has developed into a popular holiday destination, especially among honeymooners, golfers and gambling fans

JEJU

0	50	100	200 km

0	25	50	100 miles

Japanese Rail System

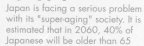

Fastest
603 km/h – world's train speed record broken in 2015

Punctual
The average train delay in Japan is just 18 seconds

Russo-Japanese War

This 1904–5 war ended with a complete Japanese victory over Russian superior strength and for the first time an Asian power defeated a European one

Aging population

Japan is facing a serious problem with its "super-aging" society. It is estimated that in 2060, 40% of Japanese will be older than 65

Kongo Gumi
For over 1,400 years, until its liquidation in 2006, it was the world's oldest private company. It specialized in building Buddhist temples

Centenarians

In 1963, when Japan started collecting data about its aging population, there were only 153 Japanese older than 100 years. In 2016, their number reached 65,692

SEA OF OKHOTSK

Daily tremors
Located at the junction of four tectonic plates, Japan is hit by around 1,500 earthquakes every year

HOKKAIDO
Similar in size to the island of Ireland

Obesity policy
The law says that when you're 40, you must go for an obligatory medical check-up and if you're overweight you must follow a strict diet and get in shape

Seikan Tunnel
Hokkaido and Honshu islands are connected by a 53-km-long underwater railway tunnel

SEA OF JAPAN (EAST SEA)

Dancing prohibition

After WWII, late-night dancing was illegal in Japan. The ban was lifted in 2015 – before then, even if a venue had a dancing licence, all dancing had to stop at midnight

Karoshi
"Overwork death". In 2015, over 2,300 people died from overwork in Japan

TASHIROJIMA
Cats on this island outnumber humans. Originally, they were supposed to chase mice away from precious silkworms

HONSHU

Greater Tokyo Area
With a population of more than 38 million and an area similar in size to Montenegro, it's the most populous metropolitan area in the world

JAPAN

★ ## 2011 Great East Japan earthquake
4th strongest recorded earthquake (9.0 magnitude). Together with the resultant tsunami, it is the costliest natural disaster with $235 billion worth of damage

Fukushima nuclear disaster
Caused by the 2011 tsunami, it is the 2nd most significant nuclear disaster after the 1986 Chernobyl disaster

"Double survivor"
One man survived both nuclear attacks during WWII. Tsutomu Yamaguchi was on a business trip in Hiroshima and returned home to Nagasaki the day before the second bombing

Oleander
Official flower of Hiroshima. It was first to bloom again after the nuclear attack

With over 1,600 temples, the city was spared nuclear bombing during WWII. Nagasaki was hit instead

Nishiyama Onsen Keiunkan
World's oldest hotel, run for the past 1,300 years by 52 generations of the same family

The Flame of Peace

Has been burning in Hiroshima since 1964 in honour of the victims and will be extinguished only when all nuclear weapons are removed from the world

90,000–146,000 casualties of the nuclear attack

KYOTO
OSAKA
HIROSHIMA

TOKYO

Tsukiji Market

World's largest wholesale seafood market

Fuji-san
The mountain's summit does not lie in any prefecture, but is considered a sacred shrine. Japanese Buddhists believe the mountain is a gateway to a different world

SHIKOKU

Tokyo-Osaka
The busiest high-speed rail line in the world. It carries over 150 million passengers per year

Seppuku
Ritual suicide, first performed during the Battle of Uji in 1180. Warriors wanted to avoid torture, shame or falling into enemy hands

KYUSHU

NAGASAKI
39,000–80,000 casualties of the nuclear attack

OKUNOSHIMA
Location of a poison gas factory during WWII. The gas was tested on rabbits which took over the island after being released into the wild

PACIFIC OCEAN

Whaling

Despite international criticism and bans, Japan's recent whaling programme aims at hunting 3,000 Antarctic minke whales over 10 years

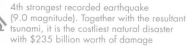

King cobra
World's longest venomous snake, up to 5.7 m in length. Its diet is based mostly on consuming other snakes. Just before its eggs start to hatch, the cobra leaves her nest permanently to avoid the temptation of eating her young

Longest conflict
World's longest ongoing civil war, consistently violating human rights with children recruited as soldiers. Around 250,000 people have been killed since 1948

Income gap
One of the highest levels of income inequality

Units of measurement
One of only three countries which has not adopted the metric system. Others are the US and Liberia

Mekong

Population

Vietnam   93,448,000
Thailand   67,959,000
Myanmar   53,897,000
Cambodia   15,578,000
Laos   6,802,000

French Indochina
French colonial territory from 1887 to 1954. During WWII it was occupied by the Japanese

Snub-nosed monkey
Discovered in 2010, it is already critically threatened. When it rains, it hides its head between its knees as raindrops make it sneeze due to its short upturned nose

MYANMAR

★ World's largest book
1,460 pages, each 1.5 m²

★ Bagan
Ancient city, founded 1,200 years ago, once the capital of the Pagan Empire. It included over 10,000 temples, of which 2,200 survive

NAY PYI TAW ●
Construction of this new capital city began in 2002

Giant huntsman spider
The spider with the largest ever leg-span (30 cm), was found in Laos in 2001

Motorbikes
There are around 4.9 M motorbikes in Hanoi

Unwanted Nobel Prize
Le Duc Tho, a Vietnamese politician, is the only person who refused a Nobel Peace Prize, in 1973. He said he might reconsider the award when peace is truly re-established in Vietnam

VIETNAM

HANOI ●

Bombs
World's most bombed country per capita. Over 2 million tonnes of bombs were dropped during the Vietnam War

LAOS

Narrow houses
Hanoi's houses are narrow because real estate tax is based on width

Saola
This small, shy mammal living in the Annam Highlands is one of the rarest animals on the planet and was discovered by a WWF team in 1992

VIENTIANE

Burma
Myanmar is also known as "Burma"

Bumblebee bat
The smallest mammal, weighting just 2 grams, is endangered and lives in caves in Myanmar and Thailand

Elephants
In the 19th century, over 100,000 elephants lived in Thailand. Today only about 5,000 remain

Rice
Thailand had been the largest rice exporter in the world until 2015, when it was surpassed by India

Monkey Buffet Festival
Held annually as a thank you for bringing thousands of tourists. Around 2,000 monkeys are fed fancy human treats

THAILAND

Vietnam War
58,315 American soldiers were killed, and on both Vietnamese side there were up to 2 millio casualties

Khone Falls
The largest waterfall in Southeast Asia

The French
The Vietnamese expelled the French in 1954 after the 7-year-long First Indochina War

THANBYUZAYAT ○

"The Bridge on the River Kwai"
This famous movie is based on the historical context of the Burma Railway construction (1942–3). Over 100,000 people died in the course of this project of which 12,621 were Allied POWs

Golden Buddha
The Golden Buddha statue is 3 m high, made of 5.5 tonnes of gold worth up to $250 million

This complex of temples is the world's largest religious monument

Khmer Rouge rule
During its 4-year rule, almost 50% of the population died of starvation, torture, execution or overwork

Angkor Wat ★

BAN ○ PONG

● BANGKOK

Tonle Sap

Cambodia's Water Festival

CAMBODIA

King Bhumibol Adulyadej
After the death of the Thai king, greatly respected by his subjects, a one-year period of national mourning was announced

PHNOM PENH ●

HO CHI MINH CITY

ANDAMAN SEA

GULF OF THAILAND

Former Saigon
Upon establishment of the Socialist Republic of Vietnam in 1975, the ci was renamed after Ho Chi Minh, the communist revolutionary leader

Mekong

Islands
Thailand includes over 1,400 islands

Ko Phangan
The famous Full Moon Party was first organized here as a small beach party by hippies in the 1980s

Mekong

Reserve military
Today, Vietnam has the largest reserve army in the world, including over 5 million reservists

Independence
Thailand is the only country in the Southeast Asia region which has never been colonized by Europeans

Cambodia's Water Festival
When the summer monsoon comes, Tonle Sap river reverses its current and floods Tonle Sap lake. Despite the destructive side of this, as the river reverses back it leaves tonnes of fish and nutrients, which fertilize the soil securing the survival of local communities

0   100   200   400 km

0   50   100   200 miles

"War on drugs"

Since Rodrigo Duterte became the president in 2016, the Philippines has become the most dangerous country to traffic or use drugs. Government-backed militias reportedly execute buyers and sellers without a trial

BATAN ISLANDS

Idjang citadels

These triangular-shaped fortifications were camouflaged to merge into the green, tropical hilltops and used by the Ivatan people from 2000 BC. They were discovered in 1994

Philippine tarsier

It's one of the smallest and oldest mammal species on Earth. It cannot move its eyeballs, which are larger than its brain, but it can turn its neck 180° to each side

SOUTH CHINA SEA

PHILIPPINES

During his 1542 expedition, Spanish explorer, Ruy López de Villalobos, named the archipelago to honour his king, Philip II of Spain. Today, it's the only Spanish-speaking nation in Asia

Banaue Rice Terraces

These rice terraces have been cultivated for the past 2,000 years, a magnificent wonder of agriculture carved, mostly by hand, into the high altitude green mountains of Ifugao

Sagada's hanging coffins

In this ancient way of burial, the deceased are placed in coffins which are suspended on cliffs above the village

Deforestation

Only around 10% of the original Philippine forests survive today

PACIFIC OCEAN

Mount Pinatubo (1,660 m)

Its 1991 eruption was the largest in recorded history to affect a densely populated area. It was so violent that the mountain collapsed in flames and was reduced in height by more than 250 m

LUZON

PHILIPPINE SEA

Greatest population density

Manila is the world's most densely populated city with 41,515 people per km² (twice that of Paris)

MANILA

Island nation

Although the Philippines is made up of 7,641 islands, 95% of the country's land area is concentrated in its 11 largest islands

Taal volcano

Its crater lake is the world's largest lake on an island in a lake on an island

"Perfect cone"

Mayon volcano has a perfect conical outline

Typhoon Haiyan

Typhoon Haiyan

One of the strongest tropical cyclones ever recorded hit the Philippines in 2013. It caused $2.86 billion of damage and left 6 million people homeless

Galathea Depth

At 10,540 m below sea level, it's the 3rd deepest place in the world's oceans

PALAWAN

Magellan's death

Killed during a battle on Mactan island in 1521, as he was trying to baptize the local population. The rest of the circumnavigation of the globe was completed without him

World's largest pearl

Pearl of Puerto is the largest pearl ever found. It weighs 34 kg and its estimated value is $100 million

BOHOL SEA

Tubbataha Reef

UNESCO World Heritage site, a coral reef with an exceptionally high density of bird and marine species. It was nominated as one of the New 7 Wonders of Nature

SULU SEA

Hikaru Sulu, a character from the famous "Star Trek" series, was named after this sea

MINDANAO

"Food basket" of the Philippines. The vast majority of the country's food is produced on this island

Philippine Trench

Mountain full of life

Mount Apo has 4 lakes, 19 rivers, a 100-m-tall waterfall and a tropical rainforest inhabited by over 100 endemic animal species and many unique plants

Mount Apo △ (2,954 m)

Volcano and the tallest mountain in the country

Largest banknotes

Measuring 21.5 cm x 36.0 cm, the 100,000-peso note is the largest of all legal tender notes in the world. Only 1,000 were issued in 1998 to commemorate 100 years of independence for the Philippines

SULU ARCHIPELAGO

Magellan's circumnavigation

CELEBES SEA

Volcanic power plants

Mount Apo features two geothermal power plants, sourcing the energy from the volcano and supplying electricity to two regions

| 75 | 150 | 300 km |

| 45 | 90 | 180 miles |

Malay Archipelago
The archipelago includes Brunei, East Malaysia, East Timor, Indonesia, Philippines and Singapore, consisting of more than 25,000 islands. It is the world's largest archipelago by area

Elective monarchy
Malaysia, Cambodia and Vatican City are currently the only true elective monarchies

SOUTH CHINA SEA

Development
Thanks to rich petroleum resources, Brunei has become the 2nd most developed country in Southeast Asia (after Singapore)

Federation
Malaysia became a federation in 1963 after the unification of Malaya, North Borneo, Sarawak and Singapore. Two years later, due to the ideological differences between the members, Singapore was expelled from the federation

Bruneian Empire
At the beginning of the 16th century, at its height, the empire had control over most of the island

Flor de la Mar
Portuguese ship that sank near here in 1511. It may have treasure worth up to $2.6 billion on board. Despite many attempts, it has never been discovered

Malay Peninsula

MALAYSIA

BRUNEI

BANDAR SERI BEGAWAN
With a population of around 27,000 people, it is one of the least populated national capitals

Taman Negara
One of the oldest tropical rainforests in the world (around 130 million years old)

Deer Cave
Around 3 million bats live in this cave

Sarawak Chamber
World's largest cave chamber by area, could house 5 rows of 8 Boeing 747

● KUALA LUMPUR
● PUTRAJAYA

Lake Toba
The largest volcanic lake in the world, it was created after the largest known eruption of a supervolcano, nearly 70,000 years ago

SAMOSIR
The largest island on an island

SINGAPORE

Ease of doing business
Singapore led the "Ease of doing business" for 11 consecutive years since this ranking was introduced in 2006. It was surpassed by New Zealand in 2017

Three countries
The island of Borneo is shared by three countries, more than any other island

BORNEO
Third largest island in the world

SUMATRA

Volcanoes
Indonesia has 139 volcanoes. Only the US and Russia have more

Endangered species
The Sumatran elephant, Sumatran orangutan, Sumatran rhinoceros and Sumatran tiger are all critically endangered species due to rainforest logging and fires

Orangutan
The name derives from Malay and Indonesian words meaning "Man of the forest". In the wild, it can only be found on Borneo and Sumatra

Coconuts
Indonesia is the world's largest coconut producer

Islam
Largest Muslim country by population with 225 million believers

World's largest individual flower
Rafflesia arnoldii weighs around 7 kg and can grow up to 105 cm in diameter. When it blooms, it releases an odour comparable to rotten meat. It is endemic to Sumatra and Borneo

INDONESIA
One of the largest archipelagos by number of islands, consisting of up to 13,466 of which 922 are inhabited

Giant eruption
The volcano Krakatoa had one of its most destructive eruptions in 1883, with over 36,000 deaths attributed to the event and the tsunamis it created. It was heard 4,800 km away

JAKARTA
Greater Jakarta metropolitan area is the world's 3rd largest (31,689,592 people)

JAVA
The most populous island on the planet

BALI

LOMBOK
SUMB

Javanese
With 82 million speakers, it is one of the most widespread languages without an official status

INDIAN OCEAN

CHRISTMAS ● ISLAND (Australia)
Named after the day of its discovery in 1643, it never had an indigenous population. Under British rule, it was used for phosphate mining, a by-product of volcanic eruptions

★ *Java Trench*
Deepest place in the Indian Ocean (-7,125 m). It is considered part of the Pacific Ring of Fire

Wallace Line

COCOS (KEELING) ISLANDS (Australia)
The islands have no native mammals: there are only 2 species of rodents, 4 species of reptiles including a blind snake, and birds

0	125	250	500 km

0	75	150	300 miles

Wallace Line
Faunal boundary line named after the naturalist Alfred Russel Wallace who noted that completely different animal species live on either side of the line

PACIFIC OCEAN

Kinabalu National Park
The first Malaysian UNESCO World Heritage site. Described as one of the world's greatest biological sites

PALAU
Consists of around 340 islands that make up an area of just 497 km²

MELEKEOK
With a population of 391, it is the smallest capital of a sovereign country

Largest leaf
The largest undivided leaf was found here in 1966. It was 1.9x3 m

Crossroads of 3 major ocean currents
A huge number of large pelagic predators, turtles, dolphins, sharks and many species of migratory fish gather around the Palau archipelago

Wallace Line

CELEBES SEA

HALMAHERA

Shortest males
Indonesia has the lowest average male height – 158 cm

MOLUCCA SEA

SERAM

SULAWESI

Cutting fingers
Women from the Dani tribe are famous for ritually cutting their own fingers to honour the deaths of their loved ones

World's shortest river
Tamborasi river is only 20 m long. It is claimed to be the shortest in the world

BANDA SEA

NEW GUINEA
5 to 10% of the world's fauna and flora species live on this island, roughly the same as in the US

Oldest cave paintings
Oldest hand stencil in the world, 39,900 years old

Komodo dragon
The largest species of lizard, up to 3 m long

Puncak Jaya (4,884 m)
The highest peak of the continent of Oceania and the world's highest island mountain

FLORES

DILI

TIMOR

EAST TIMOR

Late independence
In 2002, East Timor became the first new sovereign state of the 21st century

ARAFURA SEA

SUMBA

Christianity
Together with the Philippines, East Timor is the only Christian country in Southeast Asia

Hobbit
In 2003, archaeologists discovered a skeleton the size of a 3-year-old child but with the teeth and bone structure of an adult. It is believed that it was a Homo erectus that fled from Africa 1 million years ago

TIMOR SEA

Tourism in Indonesia
Countries with the highest number of tourists visiting Indonesia per year: Singapore (1.6 million), Malaysia (1.4 million), China (1.2 million) and Australia (1.1 million)

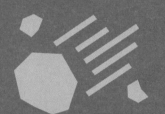

Disputed meteorite impact

There is a hypothesis that the surface of the western Central African Republic resembles a massive crater, with the Bangui anomaly in the centre and carbonados (black diamonds) scattered in the region. If true, it would be the largest known impact crater

AFRICA

AFRICA

Area
At 30,343,578 km², Africa includes over 20% of the Earth's land area. It is larger than China, India, the US and most of continental Europe combined

Population
Africa's population is expected to multiply nearly 4 times by the end of this century – it's the only continent forecast to grow as rapidly

Africa (16.36%)
1,216,130,000 people

Youngest continent
In Africa, the median age of the population is around 20 years old, while in Europe it's more than 40 years old

Wealth
Africa is expected to be the fastest growing market in the 21st century. However, the continent still includes roughly 3/4 of the poorest nations on the planet with a nominal GDP of $3.3 trillion, slightly less than Germany's, constituting just 4% of the world's total GDP

Most countries
Of all the continents, Africa has the greatest number of countries, including 54 sovereign states, and two states with limited recognition – Western Sahara and Somaliland

Africa's challenges
Africa's greatest challenges focus around booming urban populations, crippling infrastructure, social inclusion, access to education and health services

• ASCENSION (UK)

Dictatorships
Currently, there are 21 countries in Africa recognized as authoritarian regimes ruled by dictators, some of whom have commited crimes against humanity

ST HELENA (UK) •

ATLANTIC OCEAN

Ras ben Sakka
Northernmost point in mainland Africa
37°21'0"N

Sahara
The largest hot desert on the planet, it has an area of 9,200,000 km² and is similar in size to the US

First of our kind
A 2017 discovery in northern Africa proves that the first humans capable of using fire and tools emerged up to 350,000 years ago, roughly 100,000 years earlier than previously thought

Africa is the longest inhabited territory on the planet

Trans-Sahara Highway
Around 4,500 km long, of which 85% is paved. Its central part requires special vehicles to cross the desert

Trans-Sahara Highway

Slave trade
The large scale slave trade to the Americas began in West Africa in the 18th century

Pointe des Almadies
Westernmost point in mainland Africa
17°33'22"W

Niger

Protected areas
There are more than 3,000 protected areas in Africa

Population growth
Nigeria (182.2 million poeple) is expected to surpass the population of the US by 2050 and become the 3rd most populous country on Earth

West Africa
Includes 17 countries. 353 million people live in this region. If counted as a separate country it would have the 3rd largest population but would only be 23rd by GDP

LAGOS
Population:
14.2 million

The first globe
In 1486, Martin Behaim, a German artist and navigator, took part in the first European expedition to enter the mouth of the Congo river. On his return, he designed the first terrestrial globe, known as Erdapfel

Phoenician sailors
According to Greek historian Herodotus, the first circumnavigation of Africa was made by Phoenician explorers around 600 BC

Portuguese colonization
Portugal was the first European power to establish settlements and trade posts during the post-Middle Ages along the coast of Africa

Scramble for Africa
Occupation of Africa by European powers between the late 18th and early 20th centuries. In 1913, only Ethiopia and Liberia were independent

Cap Agulho
Southernmost po[...] in mainland Afric[...]
34°50'0[...]

| 0 | 500 | 1,000 | | 2,000 km |
| 0 | 300 | 600 | | 1,200 miles |

Suez Canal
Construction of the Suez Canal, completed in 1869, played an important role in, and accelerated the process of, European colonization of Africa

Dark skin
After early hominid species moved from the jungle to sunny savannas around 1.2 million years ago, their skin needed an improved cooling mechanism. This was achieved by the loss of body hair which eventually led to dark skin pigmentation as a result of ultraviolet radiation

CAIRO
Population: 19.5 million

Septimius Severus
21st Emperor of Rome (AD 193–211), born in Africa in what is now Libya

Sinai peninsula
The only land bridge between Africa and the rest of the world after the Gibraltar Arc closed around 6 million years ago

Malaria
Around 75% of malaria cases occur in 13 countries of Sub-Saharan Africa. It's a life-threatening disease carried by a parasite and transmitted by a mosquito bite

Nile

Bridge of the Horns
There is a plan to build a 29-km-long bridge over the Bab al Mandab which would become the world's longest suspension bridge

Hunger
Almost a quarter of the population (around 250 million people) is undernourished. Up to 40% of them are children under 5 years old

Child labour
Over 40% of children in Africa aged 5–14 are engaged in child labour

Water
People in Africa have to walk an average of 6 km to collect water

Blue Nile

Raas Xaafuun
Easternmost point in mainland Africa 51°27' 52''E

Lake Assal, Djibouti
The lowest point in Africa (-156 m)

Life expectancy
The average life expectancy in Africa is 60 years, compared to 71 worldwide

Gorillas
The largest living primates. Between 95% and 99% of their DNA is similar to that of humans

White Nile

Colonization
Besides Ethiopia and Liberia, every African nation was once colonized

First humans
According to the popular "Recent African origin of modern humans" hypothesis, anatomically, modern humans evolved in East Africa around 100,000 – 200,000 years ago

Geographic centre
Lobeke National Park in Cameroon is the geographic centre of the continent

Running
Many of the world's fastest long-distance runners come from a single ethnic group in Kenya, called the Kalenjin

Overfishing
Chinese distant-water fishing vessels exploit the waters of African countries. The hand-hewn boats of local fishermen can't compete with mega-trawlers

Rainforest trees
Tropical rainforest trees are so densely packed that rain falling on the canopy can take as long as 10 minutes to reach the ground

KINSHASA
Population: 12.6 million

Congo

David Livingstone
This Scottish explorer led an expedition to find the source of the Nile. He was considered a British "hero" and thought that if he found the Nile's source, he would gain the power and popularity to influence and end the East African Arab-Swahili slave trade

Greatest animals
Africa is home to the world's largest land animal, the African bush elephant, and the tallest one, the Masai giraffe

Poorest nations, rich in resources
Africa holds around 30% of the Earth's mineral resources. They generate over 40% of the continent's public revenues and account for over 70% of exports

Maps remembered from school
Africa is the largest victim of the Mercator map projection, as it distorts the relative sizes of countries by stretching the lines of latitutde as you go further from the equator, making countries in the high latitudes look larger than they really are

Languages
It is estimated that there are over 3,000 languages spoken in Africa which makes it the most linguistically diverse continent

INDIAN OCEAN

Tourism in Africa
Most popular destinations (visitors annually):
Morocco – 10 million
South Africa – 9.5 million
Tunisia – 6.2 million

African Union
The union of 55 African states was established in 2001 in Ethiopia to defend independence and encourage political, economic and territorial integration

Financial aid
African countries were granted over $650 billion of financial aid between 1960 and 2006. Only a few countries have successfully used this money and achieved sustainable growth

ATLANTIC
OCEAN

Spanish cities
The two Spanish administered
cities in mainland Africa

City of spies
The city gained fame as an
international espionage hub
due to its cosmopolitan status
in the 20th century

TANGIER
The oldest city in Morocco,
established 2,500 years
ago by Carthaginians

CEUTA

ME

CHEFCHAOUENE
Beautiful, small town known for its
picturesque light-blue city walls and
nearby plantations of cannabis

RABAT

Conference of 1943
Winston Churchill and Franklin D. Roosevelt
met here to plan the strategy for ending
WWII

Roosevelt's flight to the conference
was the 1st presidential airborne
journey in history

MADEIRA (Port.)

CASABLANCA

Morocco is the world'
largest producer of
peppermint, cannabis
and canned sardines

Islam
World's westernmost
Muslim country

MOROCCO

33-year split from African Union
Morocco was the only African state not to join
the African Union as it had left the Organization
of African Unity (precursor to the AU) in 1984
due to territorial conflict over Western Sahara.
It rejoined the AU at the beginning of 2017

MARRAKESH

Toubkal
4,167 m
Highest peak
of the Atlas
Mountains

*Marathon
des Sables*
This six-day ultramarathon
of around 250 km
is considered the
most difficult run
in the world

Venus of Tan-Tan
300,000–500,000 years old,
this stone object may be the
earliest artistic representation
of the human figure

CANARY ISLANDS (Sp.)

Mourning in white
A Moroccan widow wears
a white hijab for 40 days
after her husband's death
to express grief

Water access
Only 22% of urban residents ha
continuous access to water.
• 34% have it only once per day
• 24% every 2nd day
• 14% every 3rd day

Cannabis
The area of cultivation of cannabis in
Morocco is estimated at 45,000
hectares. This corresponds to the
surface area of this square

*Moroccan
administration*
The capital of Western
Sahara is controlled and
administered by Morocco

LAAYOUNE
Nearly 40% of the
0.5 million population
lives here

Hospitality
In Algeria, hospitality is a national
custom and according to the
tradition, guests are greeted with
milk and sweet dates

*Moroccan
Berm*

Population density
One of the most sparsely
populated states on the planet
with just 2 people per km²

"Unresolved state"
Currently, Western Sahara
is not an independent state,
but a disputed territory with
non-self-governing status

Polisario Front
Rebel movement of the Sahrawi,
indigenous people of western parts
of the Sahara, aiming to establish an
independent state in the region

WESTERN
SAHARA

Moroccan Berm
Made mostly of sand or stone. Its construction
started in 1980 and separates two sides of
the territorial conflict. Its total length is
2,700 km – equivalent to the straight line
distance from Moscow to Marseille

Vast, empty spaces
The area of Western Sahara is
slightly larger than that of the
UK, but its population is not
even 1% of the latter

It is the largest and most populated
disputed territory in the world

Line of control
Area
controlled by
Morocco

Area
controlled by
the Polisario
Front

Visibility restricted by
the harmattan haze
60% of the time

It costs $3 to desalinate 1 m³
of water here, but thanks to
subsidies it's sold at just
3 cents to the population

No permanent streams or
lakes, so all water must be
desalinated

Without the $800 million
subsidy programme from
Morocco, the territory would
not be viable economically

*Harsh conditions
in Western Sahara*

No arable land, so
it is necessary to
import food

Extreme temperatures with
averages reaching 40 °C in
summer and 20 °C in winter

0	150	300	600 km

0	75	150	300 miles

MEDITERRANEAN SEA

ALGIERS

TUNIS

Chréa National Park
One of the few snow skiing resorts in Africa. It is home to forests of ancient Atlas cedars, where a small population of endangered Barbary macaque still live

Coastline
The Sahara desert covers over 80% of Algeria and over 90% of the 39.7 million population live along the coast

Justice system
It is quite uncommon in a predominantly Muslim state for women to make up 70% of lawyers and 60% of judges

Mosque of Kairouan
The oldest Muslim place of worship in Africa and 4th holiest site in Islam. 7 trips to Kairouan were worth 1 hajj to Mecca

Carthage
The last capital of Phoenicia and the centre of Carthaginian civilization. It was destroyed in 146 BC by the Roman Empire

TUNISIA

JERBA
In Homer's Odyssey, sailors who came to this island and tried the narcotic lotus flower, never wanted to return home

City of a Thousand Domes
El Oued oasis is a desert city watered by an underground river. Its dome-roofed houses are home to more than 130,000 people

Chott el Jerid
Seasonal lake and the largest salt flat in the Sahara. There was a plan to create a Sahara Sea by channelling it to the Mediterranean. The project failed due to lack of funding, but it inspired the creation of the Suez Canal

Snow in the Sahara
February 1979 was the first time in recorded history that low altitude Sahara regions received snow

The next such occurrence took place in January 2017, at Aïn Sefra, with over 1 m of snowfall

Grand Erg Oriental
One of the largest sand seas on the planet, roughly the size of Greece. Dunes here are on average 120 m tall

Arab Spring
Started here in 2010. Tunisia is the only state which participated in the revolution and managed to establish democracy in the aftermath

Trans-Sahara Highway

Trans-Sahara Highway
Proposed in 1962, the first highway to cross the Sahara and one of the oldest international highways in Africa

ALGERIA

Oil
Algeria is one of the 4 African members of OPEC, with the 3rd largest proven oil reserves on the continent (after Libya and Nigeria)

Sand dunes
The icon of the Sahara, the characteristic sand dunes cover only 15% of the desert area, while the majority of the surface is a rocky plateau

In Salah
Major oasis, once an important point on the trans-Saharan trade routes. More than 200,000 date palms are grown here

There is a dune here which moves 1m every 5 years, gradually cutting the oasis in half

Africa's largest country
Algeria became Africa's largest country after South Sudan split from Sudan in 2011

Size of the Sahara
The Sahara covers an area the size of the US, or 1/3 of the entire African continent

Sahara wet period
Over 15,000 prehistoric rock paintings from around 10,000 BC are found here. They depict people hunting and fishing, which suggests that the Sahara used to be a fertile savanna

Berber
The word Berber means "free man" or "noble man". The Berbers are an ethnic group consisting of over 30 million nomads scattered mostly across North Africa

Berbers still constitute about 17% and 60% of the populations of Algeria and Morocco respectively

Military
Algeria has the largest military budget in Africa

"Magic mushrooms"
Neolithic cultures that lived here during Sahara's "wet period" may have used psychedelic mushrooms as part of their religious rituals

Ahaggar National Park
Home to the last 250 white-coated Saharan cheetahs on the planet

△ Mount Tahat 2,918 m
The highest peak in Algeria

Zidane
One of the most famous celebrities of Algerian-Berber origin is the French football player, Zinedine Zidane

Simoom
It's a hot, extremely strong, dust-carrying Saharan wind. Its temperature may exceed 54°C with humidity falling below 10%. People and animals exposed to the simoom risk death from suffocation or heat stroke

Trans-Sahara Highway

LAMPEDUSA

MALTA
SICILY

Tragic migration routes
During the past decade, over 1.5 million people have risked their lives to cross the Mediterranean Sea, hoping to find a better life

Since the 2011 Arab Spring, over 3,000 people have drowned in the Mediterranean every year trying to cross it. It's as if two passenger liners the size of the Titanic have perished every year

ZUWĀRAH   TRIPOLI

Oil fields	·
Oil pipelines	···
Water pipelines	---

BENGHAZI

National flag
During Gaddafi's rule (1969–2011), Libya had a solid green flag. It was the only single-coloured national flag in the world

Atiq Mosque
An unusual mudbrick and limestone structure. The oldest mosque in the Libyan Sahara region

GHĀDAMIS
Known as "Pearl of the desert", an ancient city, famed for its Old Town

Libya's oil
The largest proven oil reserves in Africa and 9th largest in the world

Ghādamis architecture
Homes here are divided traditionally where rooftop terraces are designated for women only, ground floors serve as storage and middle floors are family rooms

Desert rose
Beautifully shaped crystals form in arid, sandy conditions, often in desert salt flats. Loose grains join together in petals which fan out in the shape of rose blossom

Tribal people
The majority of Libya's population is divided into about twenty tribal groups

After the fall of Gaddafi in 2011, lack of unity among the tribes has become one of the reasons for the ongoing civil war

Libyan desert glass
Precious glass substance found only in the Great Sand Sea, most likely originating during a meteoric impact 26 million years ago. Used since the Pleistocene to make tools, and later became the famous element of Pharaohs' jewellery

Deserted areas
Up to 90% of Libya's land is covered by desert, the most in the world

LIBYA
4th largest country in Africa and 16th in the world, similar in size to Iran

The Great Man-made River Project
World's largest irrigation project. 2,800 km of pipes and over 1,300 wells, most over 500 m deep

Daily, it supplies over 6.5 million m³ of fresh water to Libya's largest cities

Underground water resources were discovered in 1953, when Libya was looking for oil

Waw an Namus
Beautiful, extinct volcano surrounded by an oasis of 3 volcanic lakes. It is encircled by a field of volcanic ash stretching over 20 km in diameter

Al Kufrah Oasis Project
One of the largest agricultural projects in Libya. Its circular, green fields are observable from space and each has a diameter of around 1km

In 2011, due to intensive exploitation of underground aquifers, water in the oasis dried out almost completely

Lack of water
Regardless of this resource, nearly 1/3 of Libya's population lack access to safe drinking water

Forests
Only 0.12% of Libya's land is forested. 3rd lowest in the world

Only 2% of Libya's land receives enough precipitation for agricultural cultivation

Sahara desert
2nd largest desert after Antarctica and the largest hot desert on the planet

Sahara's "wet period", which was brought on by changes in Earth's orbit, ended approximately 4,000 years ago. It once was a land of green savanna

Nubian Sandstone Aquifer
Beneath the largest hot desert in the world lies the largest known system of fossil water which accumulated during the last ice age. Its size is estimated at over twice the volume of the Caspian Sea

| 0 | 100 | 200 | 400 km |
| 0 | 50 | 100 | 200 miles |

MEDITERRANEAN
SEA

Suez Canal
Ferdinand de Lesseps led the team that built the canal which opened in 1869. It is 193 km long and reduces the journey from the Mediterranean to Asia by 7,000 km

Later he attempted building the Panama Canal together with G. Eiffel, but the project failed – thousands died and both creators were convicted of fraud

Heracleion
Ancient city, currently underwater, originally built on adjoining islands in the Nile Delta

It was discovered as recently as 2000, 2.5 km off the coast, 10 m below sea level

Nile Delta
Area: Similar in size to the Netherlands

Population: 39 million

Agriculturally one of the most fertile regions in Africa, where people have cultivated the land for millennia

Lighthouse of Alexandria
120 – 137 m tall, it was one of the Wonders of the Ancient World, destroyed in 1323 by an earthquake. It was rediscovered in 1968 by a UNESCO marine expedition

Early canals
The first canal from the Nile was probably built around 3,800 years ago

Napoleon considered building a canal in 1798 but falsely believed that the Red Sea lay 10 m higher than the Mediterranean

ALEXANDRIA

Suez Canal

The Exodus

Sphinx of Giza
World's largest single-stone statue. Its exact purpose, builder and time of construction are a mystery

CAIRO

The Exodus
The route of the exodus of up to 2 million Israelites from Egypt

Great Pyramid of Khufu
138 m tall (higher than a 40-storey building). For over 3,800 years, until AD 1311, it was the world's tallest building. It is the last survivor of the 7 Wonders of the Ancient World. Made of 2.3 million perfectly carved blocks, it is believed it was built in just 20 years

Anwar Sadat
The president of Egypt at the time the country became the 1st Arab nation to make peace with Israel, in 1979

Sadat won the Nobel Peace Prize, but was assassinated in 1981 and Egypt was suspended from the Arab League

AL FAYYŪM
One of Egypt's oldest cities, it used to be one of the most important, strategically located in an oasis stretching for 60 km. When looking from space, the city stands out as a green spot surrounded by sand

Advanced engineering
The cardinal points of the Giza pyramids are aligned with the Orion constellation and the Great Pyramid points true north

The base of the pyramid is laid out horizontally with just 15 mm error. Achieving similar precision, even with modern-day technology, would be a great challenge

Great Sand Sea

Copts
Between the 4th and 6th centuries AD, Christianity was a dominant religion in Egypt. Today, Coptic Christians make up 10% of its population

Mount Sinai (2,285 m)
According to the Bible, Moses received the Ten Commandments here

EGYPT

Nile

LUXOR
The statue of Ramses II's head is perfectly symmetrical thanks to an ancient carving technique which remains unexplained

Golden Ratio
The first modern era calculation of the Golden Ratio was achieved in 1597 by a German astronomer, Michael Maestlin

4,000 years ago, the ancient Egyptians already used the Golden Ratio and Pi in the design of pyramids

New Valley Project
Long-term plan aimed at irrigating parts of the Western Desert, transforming it into agricultural land. Half of the land involved in the project will be granted to Egyptian college graduates

ASWAN
The heaviest blocks of the Great Pyramid (up to 80 tonnes) were transported from here (over 800 km)

Forests
Only 0.07% of Egypt's land is forested. Only Qatar has less

Toshka Lakes
The lakes were created at the end of the 1990s, from the excess water gathered in Lake Nasser

Lake Nasser
One of the largest man-made lakes in the world, created by damming the Nile with the Aswan High Dam in 1970

Damming the Nile made it possible to control the annual flooding, increase agricultural production and produce electricity for many isolated villages in Egypt

However, the dam stops almost all fertilizing sediments. Farmers have to use artificial fertilizers, imparing the water quality, and causing environmental and health problems

Halaib Triangle
Disputed territory between Egypt and Sudan, currently administered by Egypt

Egyptian solar calendar
5,000 years old, it was the first 365-day calendar, similar to the one we use today

It distinguished just three seasons in accordance with the cycles of the Nile: flood, emergence and low water

Nile

Nile

Dakar Rally

The route of the 2007 Dakar Rally, the last to be held in Africa. In 2008 it was moved to South America due to security issues in Mauritania

Overfishing

Mauritania has access to some of the world's richest fishing grounds which could solve the country's problem of food security. Instead, EU vessels are heavily overfishing these waters. One European vessel catches as much fish in one day as the entire Mauritanian fishing fleet in one year

Iron ore

Mauritania is the 2nd largest iron ore producer in Africa (after South Africa)

Nomadic culture

It is difficult to estimate the exact population of Mauritanian cities as most of the population live a nomadic, migratory lifestyle

ATLANTIC OCEAN

Leblouh

In Mauritania, men consider women's obesity highly attractive

Leblouh is a brutal practice of force-feeding young girls with over 10,000 calories/day of fatty food

Most mothers claim that sending their daughters to "fatteners" is the only way to secure their futures

Farming in the desert

Only 0.5% of the country's soil is fertile, yet most of its population survive on subsistence farming

Children

3rd highest fertility rate, over 6 children per woman

2nd highest infant mortality rate, over 10%

Over 47% of the population is aged 0–14, while only 3% reaches the age of 65

The Eye of Africa

Circular geological feature, 40 km in diameter, visible from space. Despite multiple theories, its origins have not been established

Nouadhibou Bay

One of the largest ship graveyards in the world with over 300 large wrecks resting on the shores

Tattoos

In the past, when a Berber girl became a woman, her face would be tattooed with geometric, tribal symbols

MAURITANIA

Desert

90% of the country is covered by desert

Sidi Yahya Gate

According to the legend, the mosque's doors have never been opened. It is thought that their opening will mark the end of the world

NOUAKCHOTT

In 1958 it was a small village. Today it's one of the largest cities in the Sahara, home to 1/3 of the country's population

The last slavery stronghold

Slavery was abolished here in 1981, but only since 2007 has it been considered a crime

Since abolition, only 1 slave-owner has been successfully brought to justice

10% to 20% of the population is still enslaved by "masters" and human traffickers

"Lac Rose"

Named after its pink waters caused by Dunaliella salina algae

The only West African country that has never experienced a coup d'état

Ruins of Koumbi Saleh

Believed to be the capital of the mighty Ghana Empire (4th–12th centuries). Very little is known about the Empire's history though it prospered largely from trans-Saharan trade

SÉNÉGAL

Sénégal

Elections

The Gambia is the only country in the world where people use stones (marbles) to cast votes during national elections

Senegambia stone circles

Over 1,000 sacred monolithic circles, up to 2,300 years old

DAKAR

Dakar Rally

Great Mosque of Djenne

Built 800 years ago, it's the largest mudbrick building in the world. Because it's highly susceptible to weather, the people of Mali hold a "festival of plastering" and renovate it before the annual monsoon

BANJUL

THE GAMBIA

The smallest country in mainland Africa

Gambia's dictator (1994–2017), Jammeh, instigated 3-day-long weekends which begin on Thursdays

The name Gambia comes from the Portuguese "cambio" which means "trade"

Gambia

Manantali Dam

An example of an inefficient, and poorly planned, built and managed dam

BAMAKO

The last rock artists

Dogon people living in the southern parts of Mali are the last group in Africa to continue the tradition of rock painting to the present day

Oysters

Oysters are the traditional food from the Gambia river

Sénégal river

Used to be called "Bambotus" which meant hippopotamus

Niger

"Harmattan" winds

The islands are a marine extension of the Sahara. A series of hot, dust-carrying winds, visible from space, occur between November and March

CAPE VERDE

Until 1980, Cape Verde and Guinea-Bissau were ruled by the same party

Discovery

These 10 volcanic islands were uninhabited until discovered by the Portuguese in 1456

Desert islands

The longest period the islands have not had rain from the southwest monsoon is 18 years. When the rains do come they are usually short and violent

PRAIA

Sahara
The desert covers large parts of 10 countries: Algeria, Chad, Egypt, Libya, Mali, Mauritania, Morocco, Niger, Sudan and Tunisia, and the disputed territory of Western Sahara. This corresponds to 30% of Africa's area

Sand dunes
Sand dunes of the Sahara can be up to 180 m high

Domestication of camels
The nomadic Berber people used domesticated camels in Africa around the 3rd century. This was crucial for trans-Saharan trade

Garamantian Road
The easiest of the ancient trans-Saharan trade routes. Caravans could average up to 1,000 camels and for 6 months of the year wander between oases to finally reach the Mediterranean

Erg Chech Meteorites
In 2007, over 100 kg of meteorites fell on the dune sea of Erg Chech and were collected by Touaregs

Uranium
Niger's income from uranium mining (former pillar of its economy) dropped by half in the late 1980s after a decrease in world demand

Food tax
After introducing taxes on basic foodstuffs in 2005, the government of Niger faced major protests and was accused of diverting aid. Around 3 million people faced hunger

Genital mutilation
20% of girls suffer from genital mutilation as no laws prohibit these procedures

Illiteracy
The most illiterate country in the world with 19% literacy rate

The richest of all time
Although today Mali is one of the poorest and least developed countries, it was once one of the richest, where gold was in abundance

Emperor Mansa Musa of Mali, who lived in the 14th century, is considered the wealthiest man who ever lived

MALI

Islam
90% of the population is Muslim

Poverty
Average monthly salary in Mali is around $125

Tree of Ténéré ★
It used to be considered the most isolated tree on earth, 400 km from any other. It had a 40-m-deep root system. A drunk driver knocked it down in 1973

Highest birth rate
44.8 births/1,000 people

Highest fertility rate
6.6 births/woman

Radio
Radio is the most important medium. TVs are too expensive for the majority of the population and the illiteracy rate is high

NIGER

poor health care · lack of sea access
poor education · **Niger's challenges** · lack of infrastructure
desertification · poverty · overpopulation

Languages
The official language of Niger is French. There are 10 other national languages

"Athens of Africa"
From the 13th–16th centuries, the world's largest university was in Timbuktu. It hosted over 20,000 students

La Dune Rose
Big sand dune surrounded by the Niger river, grasses and occasionally, lotus blossoms. At dusk, the light reflected from the river turns the dune's colour to pink

Religion
There is no state religion. Formation of political parties based on religious doctrines is illegal

Agriculture
Most of the land is desert. Only 20% of the southern region is savanna, which is suitable for livestock and limited agriculture. It is threatened by desertification

Independence
Independence from France in 1960. Until 1991, ruled by military governments

Bandiagara Escarpment
Beautiful, 150-km-long, 500-m-tall sandstone cliff. At its base, there are many clay houses built by the Tellem people who first inhabited the area 1,000 years ago

Poverty
Majority of the population live on less than a $1 a day

NIAMEY

Co-administration
From 1932 to 1947, for budget reasons, Niger was administered together with what is now Burkina Faso

CFA franc
West African & Central African CFA francs are two main currencies used in 14 countries in West and Central Africa

Although theoretically separate, the two CFA currencies are effectively interchangeable

They're guaranteed by the French treasury and exchanged at a fixed rate to the euro

Niger river
3rd longest river in Africa. Every year between September and May it floods, bringing benefits for agriculture and fisheries, but also carrying a threat to the population

Niger

Garamantian Road

| 0 | 150 | 300 | 600 km |
| 0 | 75 | 150 | 300 miles |

Agricultural fires

Fires used to clear land and forests for agriculture are a significant environmental problem impacting weather, human health and natural resources

"Colourful rivers"

In West African Manding languages:
Baoulé – "red river"
Bafing – "black river"
Bakoy – "white river"

Sénégal

Agricultural zones

Guinea-Bissau is divided into 3 regions according to the water requirements of major crops: the coconut palm tree zone, the intermediary marshy rice zone and the sandy interior areas with peanut cultivation

Cashew nuts

World's 6th largest producer. Most of Guinea-Bissau's population depend on it

Bakoy

Baoulé

Niger

The name of the capital was added to the country's name to avoid confusion with its neighbour

BISSAU

GUINEA-BISSAU

Slavers' hideout

After the slave trade was abolished in the 19th century, many slavers, pursued by the British Royal Navy, went into hiding in Guinea

Aluminium

World's 2nd largest producer of bauxite, the principal ore of aluminium

Muslims

85% of the population are Muslim

Bafing

GUINEA

Niger

Originally built on a small island which was later connected to the mainland by a causeway as the city expanded

CONAKRY

Pygmy hippopotamus

Relatively small hippopotamid that lives only in West African forests and swamps, it is one of only two endangered Hippopotamidae species

Creole people

The descendants of freed African-American and West Indian slaves, Creoles, represent about 5% of the population

Ebola epidemic

Killed around 11,000 people across the world (2013–15) and was the deadliest occurance of the disease since its discovery in 1976

Deforestation

Since independence in 1960, the forested area of the country has decreased from 160,000 km² to 100,000 km²

SIERRA LEONE

(In Portuguese: "lion mountain range")

"Star of Sierra Leone" ★

The largest alluvial diamond ever discovered (968.9 carats) and the 4th largest gem-quality diamond in the world, found in Koidu in 1972

★ Ebola's ground zero

CÔT
(IVO

Refugees

Because of the 1991–2002 civil war, around 400,000 refugees left Sierra Leone for The Gambia, Guinea and Liberia

FREETOWN
Founded in 1792, as a home for formerly enslaved Africans

Biodiversity

The highest level of biodiversity in West Africa. More than 1,200 animal and 4,700 plant species

Diamond smuggling

Sierra Leone's legitimate economy is undermined and destabilized by rebel-controlled areas of diamond smuggling

Wasted land

Agriculture intensification, caused by an 80% population increase during the second half of the 20th century, resulted in soil depletion and deforestation

Water pollution

Caused by the mining industry, it is a serious humanitarian problem

Liberian mongoose

Threatened due to hunting, mining and pesticides that accumulate in the worms it eats

LIBERIA

Emerald cuckoo

"The most beautiful bird of Africa" has disappeared from most of West Africa but can still be found in Sierra Leone

Freed slaves

Between 1822 and 1847, 3,000 Caribbean slaves, who were freed or managed to escape, came to Liberia to establish a free country

MONROVIA

Forests

Liberia is home to great evergreen forests. Around 240 species of trees can be found here, including the wild rubber tree, mahogany, ironwood, the cotton tree and the oil and raffia palms

Coffee and cocoa production boomed between the 1960s and 1980s

Internal conflicts

Since the 1990s, the civil war and internal conflicts have killed over 250,000 and forced almost 1 million people to emigrate. The war also caused serious economic stagnation

ATLANTIC
OCEAN

Independence

During the colonization era, Liberia was the first independent country in Africa (1847)

Ellen Johnson Sirleaf

Africa's first female president. Won the Nobel Peace Price for the struggle for the safety of women and for their rights, as well as transforming the country after Charles Taylor's reign of terror

0	50	100		200 km

0	30	60		120 miles

Refugees
Around 34,000 refugees from Mali live in Burkina Faso

TAMBAO ★
Manganese mine, rich in up to 100 million tonnes of this metal

Mine railway
Railroad from Ouagadougou to Tambao, to exploit the minerals, was constructed mainly by volunteers

Mine railway

Cotton
Africa's largest producer of cotton

Gold
Main export resource of Burkina Faso

BURKINA FASO

Niger

Child labour
According to the US Department of Labor, gold and cotton are produced using child and forced labour here

● OUAGADOUGOU
The name of the capital means "You are welcome here at home with us"

Agriculture
90% of the population is engaged in subsistence agriculture

Black Volta

Deforestation
Dependence on wood for trade and fuel uses about 320 km² of forests every year

Anthem
The national anthem starts with the words "Against the humiliating bondage of a thousand years"

Dahomey Amazons
They were the all-female regiment of King Houegbadja of Dahomey (17th–19th century kingdom in what is now Benin). As an army, first used in the defeat of the Savi kingdom in 1727. The warriors referred to themselves as ahosi – "king's wives" or "our mothers"

Water supply problem
Burkina Faso has up to 13 billion m³ of improved water. 82% of urban and only 44% of rural populations have access to safe water

WHO claims that approximately 80% of all diseases in the country are caused by poor water. It is a result of uncontrolled disposal of sewage and industrial waste

"Witch camps"
There still exist small villages in the north, populated mostly by victimized older women, believed to be witches

White Volta

Soviet allies
Benin and the Soviet Union became major political allies after Mathieu Kérékou came to power in a coup in 1972 and started following a Marxist-Leninist course

Fuel trade
One of 5 African countries that banned the import of high-on-sulphur fuels from European companies that exploit the weak regulations of West Africa

BENIN

Transport
One of the most developed transportation systems in Africa

Labour force
In 1988, 28% of workers were foreigners because of the country's flourishing economy since gaining independence. Around 1 million migrants from Burkina Faso worked on cocoa and coffee plantations here

Gold Coast
The name of the British colony from 1821 to 1957

The word Ghana means "warrior king"

Pel's flying squirrel
As with other species, it is endangered

TOGO

Voodoo
Birth place of African Vodun, better known in popular culture as Voodoo

Togolese ethnic groups
About 40 tribes, with distinct languages and cultures, are found in Togo. The largest ethnic group is Ewe in the south (2.1 million people)

IVOIRE
(AST)

African elephant
The largest land animal on Earth (6 tonnes). Due to ivory poaching, its population has declined by over 90% (from 5 million to 415,000) during the last 70 years

GHANA

Kofi Annan
This Ghanaian, former Secretary-General of the UN, co-received the Nobel Peace Prize in 2001 for fighting for human rights

Lake Volta
Largest man-made lake in the world by area

Slave trade
Atlantic slave trade began here in the 16th century

African elephant, Diana monkey, chimpanzee and red-bellied monkey are endangered species in Togo due to deforestation and poaching

Gold
Recently, Benin liberalized its gold mining law to attract foreign explorations and investors

● YAMOUSSOUKRO
Country's 5th most populous city

Exports

Today
Coffee, cocoa, tropical wood, palm oil, rubber, cotton, bananas, pineapples

16th century
Ivory, arabic gum, pepper, gold, slaves

Akosombo Dam
Major investment to provide electricity for the aluminium industry. It caused displacement of many people and environmental changes after Lake Volta was created

PORTO-NOVO

LOMÉ
Founded in the 19th century by European traders

Slave Coast

Zetahil
Religion that combines elements of both Islam and Christianity. It is unique to Ghana

ACCRA

ABIDJAN
West Africa's largest container port and the largest tuna fishing port in Africa

Gold Coast

Ivory Coast

The largest slave market until the 19th century

Slave Coast

6 million people were sold and shipped to Asian markets

12 million people were enslaved and shipped across the Atlantic

4 million people were killed or died on the continent after capture

Ivory Coast
In 1989 the international trade of ivory was banned at the Convention on International Trade in Endangered Species of Wild Fauna and Flora

Gold Coast
European colonists found gold here in the 16th century

Today, one of the largest transshipment points of drugs moved from South America to Europe

Ghana is still the 2nd largest gold producer in Africa and gold is Ghana's main export product

Youth population
With a population of 33 million people aged 18–35 years, Nigeria has one of the largest youth populations in the world

Cassava
World's largest producer of cassava. It is a root plant and major part of the diet of over 0.5 billion people living in tropical climates

Traditional workshops
In 1983, when the economy almost collapsed, most people could no longer afford factory-made or imported products. Traditional workshops (soap, cane, wood, etc.) became widespread in rural areas

First free election
After years of military dictatorships, the 2011 presidential election was considered the first to be fair and free in Nigeria's history

Former British colonies

Oil boom
Nigeria experienced an oil boom in the 1970s which significantly developed the economy

Savanna
The northern half of the country is covered with short and tall grass savanna

Zuma Rock
Large, 725-m-tall, single massive rock, rising steeply out of its surroundings

Kainji Dam
Built between 1964 and 1968 for approximately $200 million. Around 3/4 of this budget was used to relocate the population

Niger

Oil spills
Many individuals steal oil by drilling holes in pipelines. When they are finished, they leave the pipe open which leads to large oil spills. It is a common practice

Sharia law

ABUJA
Planned city, mostly built in the 1980s

NIGERIA

Oil and gas industry
This industry contributes approximately 35% to Nigeria's GDP

Benue

Economy
The largest African economy and 26th in the world with a nominal GDP of $415 billion. At this rank, it is slightly above Iran and below Poland

Butterflies
These lands contain one of the world's largest diversity of butterflies

Ethnic groups
Over 500 ethnic groups speak over 500 languages

Niger

Most populated city in Africa (14.2 million people)

Next Eleven
Goldman Sachs identified 11 countries as those with a potential to become a superpower: Bangladesh, Egypt, Indonesia, Iran, Mexico, Nigeria, Pakistan, Philippines, Turkey, South Korea and Vietnam

LAGOS

Civil war (1967–1970)
Started after the Republic of Biafra declared independence in 1967. 1–3 million people died from gunfire, diseases or starvation before the Nigerian government regained its lost territory

Republic of Biafra

Conflict in the Niger Delta
Lasting from 2003 to the present day, it is a conflict between the local native population and international oil corporations who exploit these lands. Illegal activities carried out by locals include piracy and kidnapping

Niger Delta
Home to 20 million people

From 1976 to 1996, 1.89 million barrels of oil were spilled here causing major environmental damage

Mt Cameroon
Active volcano and the highest point in Cameroon (4,100 m)

GULF OF GUINEA

Against apartheid
The Nigerian government played an important role in fighting apartheid in South Africa

0	100	200	400 km
0	50	100	200 miles

Boko Haram

Islamic terrorist organization that originated in Nigeria and is also active in Chad and Cameroon. Responsible for over 50% of all fatalities in the world caused by acts of terror in 2014

Schoolgirls kidnapping

In 2014, Boko Haram kidnapped 276 girls from a secondary school in Chibok. 112 are still missing

The greatest extent of Boko Haram, in 2015

Sharia law

Nigerian states that lie north of this line have Sharia law

Sharia law

Benue

Last rhino

The last Western black rhino lived in northern Cameroon. It was declared extinct in 2011

English

English is the official language of Nigeria as it is a former British colony

Benue river

This is a navigable river and an important transport route for both countries, but only during summer months

Paul Biya

Current president of Cameroon, in office since 1982

Population

Nigeria has the highest population in Africa and 7th in the world (182.2 million). Since 1971, it has increased nearly 3.5 times

Colonization

Cameroon is one of four former German colonial areas

Exploding lake

In 1986, a buildup of CO_2 in Lake Nyos erupted, suffocating thousands of people and animals

Only 3 such lakes exist in the world: 2 in Cameroon, 1 in Dem. Rep. Congo

CO_2

Adamaoua region

Mountainous region that divides the country between savanna north and forested south. It is often called the "water tower" of Cameroon because many rivers start here

Drill

These primates can live in groups of over 100 individuals. They are found only on the border of these two countries

CAMEROON

Known as "Africa in miniature" for its geographical diversity

Wouri river

Named "Rio dos Camarões" by early Portuguese explorers, meaning "Shrimp river". The name of the country was formed from this

Wouri

DOUALA

Cameroon's largest city (2.2 million people)

YAOUNDÉ

The population of Yaoundé is strongly engaged in urban agriculture

Unemployment rate

Formally, the unemployment rate in Cameroon is only 4% but the real rate may be as high as 30%

Languages

Around 280 indigenous linguistic groups live here, while English and French are still official languages

Football

One of the most successful football teams on the continent. Played 7 times at the World Cup (more than any other African country) and the first African team to reach the quarter-finals

Religion in Nigeria

The Christian population lives mainly in the southern regions of the country and Muslim, in the northern

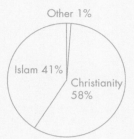

Other 1%

Islam 41%

Christianity 58%

Former German colonies

Development

Last in the Network Readiness Index, which assesses the development of a country's information and communication technologies

Christianity

Over 50% of Chadian Christians practice polygamy and approximately 40% of all women live in such unions

"Sailing through the Sahara"

In 1851, a boat was carried across the Sahara by camel to complete the 1st scientific waterborne survey of Lake Chad

Emi Koussi (3,415 m) △

Volcano, the highest mountain in the Sahara

Earliest common link

★ 7 million-year-old fossils discovered in 2001 are the oldest known remains of a common ancestor of humans and chimpanzees

Bodélé depression dust storms

They take place on average 100 days per year and can stretch from here to Cape Verde. They carry 700,000 tonnes of dust each day, the mass of two Empire State Buildings

"Ashes to Ashes..."

This area is extremely rich in nutrients as it used to be under the waters of Lake Mega-Chad. These nutrients are carried by dust storms over the Atlantic to the Amazon Basin and are crucial for its growth

Bodélé depression

Guelta d'Archei ★

One of the most famous and beautiful oases. Several kinds of species live there, including the last colony of the Nile crocodile in the Sahara. To get there, one has to travel at least 4 days from Chad's capital

Lake Mega-Chad

In 5000 BC, Lake Chad was the world's largest lake (1 million km²), more than twice as large as the Caspian Sea of today

Lake Mega-Chad

CHAD

Infrastructure

Decades of civil war have crippled Chad's development and infrastructure. It's about twice the size of France and in 2000 it had roughly:

Darfur genocide

The first genocide of the 21st century, carried out in 2003 by Omar al-Bashir's government. It is estimated that 300,000 people were killed and up to 3 million were displaced

Lake Chad in the 1960s

Around 10,000 telephones and just 1,000 internet subscribers

1 television set per 1,000 citizens, with just one TV station available

600 km of paved roads and 5,600 km of dirt roads obstructed during rainy season

Omar al-Bashir

The 1st sitting president in history to be indicted by the International Criminal Court for mass killing: He is still in power

NDJAMENA
Constant water and energy services are provided only in the capital

Lake Chad today

Since the 1960s, Lake Chad has shrunk by more than 90%. Its current surface area is just 1,350 km² and its average depth only 1.5 m

Children of Chad

In 2014, 30% of children aged 5–14 (1 million) were working, 18% aged 7–14 combined work and school, and only 35% attending school completed primary education

Light pollution

One of the best countries to see the night sky because it's little affected by light pollution

Scarce resource

Lake Chad is a source of water for over 68 million people

Natural wealth

The country is rich in resources such as uranium, cobalt, crude oil, gold and diamonds, but it remains one of the poorest on the planet

"Diamond Affair"

In 1973, dictator Jean-Bédel Bokassa offered the French Minister of Finance, Giscard d'Estaing, two diamonds worth a fortune. It sparked an international scandal and contributed to Giscard d'Estaing losing his 198 reelection bid for presidency

CENTRAL AFRICAN REPUBLIC

Blood wood

In 2013, European corporations paid over 3.4 million euros to guerilla fighters to benefit from illegal logging of the rainforests

CAR has received millions of euros in development aid from the EU, based on the flawed assumption that CAR's logging industry contributes to local development

Disputed meteorite impact

There is a hypothesis that the surface of western CAR resembles a massive crater, with the Bangui anomaly in the centre and carbonados scattered in the region. If true, it would be the largest known impact crater

Carbonado

This "Black diamond" is the toughest mineral on Earth. It's older than Earth which means that it fell from space in a meteorite. It can only be found in CAR and Bahia, Brazil

Gravity anomaly

Largest negative gravity anomaly – things weigh less and compasses fail to point north. Its origin is unknown

BANGUI

Ubangi

Uele

Ubangi-Uele

The combined length of these rivers is about 2,270 km. In the 1960s, there was a plan to divert them so that they emptied into Lake Chad

Diversity
Country of about 40 million citizens, divided into 597 ethnic groups and tribes, speaking over 400 different languages and dialects

Sharia law
Sudan's legal system is based on Sharia law. Flogging, stoning to death and crucifixion remain legal forms of punishment

Osama bin Laden
He was exiled from Saudi Arabia in 1991 and bought a house in Khartoum where he lived and ran multiple businesses. In 1996, he declared war on the US and returned to Afghanistan

SUDAN

White Nile
Originates in the African Great Lakes. It is one of two main tributaries of the Nile and adds an extra 3,700 km to the length of the Nile proper, making the Nile one of the two longest rivers in the world

White Nile conquest
In 2004, Hendrik Coetzee became the first person to navigate the entire course of the Nile from its source to the Mediterranean. He died on an expedition in 2010, eaten by a crocodile in the Congo river

Oil
Most of former Sudan's oil fields now belong to South Sudan

Abyei
Oil-rich district, inhabited by the Ngok Dinka tribe. Both Sudan and South Sudan claim it as their own

The hat
Since South Sudan's president, Salva Kiir Mayardit, received a stetson hat from the US president George W. Bush, he makes every public appearance wearing it

SOUTH SUDAN

Civil war
Since 2013, this newly formed country has been plunged into a civil war that has already claimed over 300,000 lives

"Failed state"
In the years 2013–14, South Sudan led the Fragile States Index which measures countries' vulnerability to collapse

JUBA

Nourishing Nile
Every year when the Nile overflowed, before it was dammed, it left a thick layer of silt rich in nutrients. Egyptians even had a god of the floods

The longest river?
In 2007, researchers found an estuary which made the Amazon nearly 150 km longer than the Nile. However, the "true source" of the latter is still debated

Secession
Before the South Sudan secession, it was the largest country in Africa by area. It's 3rd now after Algeria and Dem. Rep. Congo

Nile

Nubian Pyramids of Meroë
It's the greatest concentration of pyramids on the planet. There are 255 of them, each one about the size of a 10-storey building. They were built by the rulers of the Kushite Kingdom, about 1,500 years after the Egyptian pyramids

KHARTOUM

White Nile

Blue Nile

Blue Nile
Supplies 59% of water to the Nile proper and 90% of the river's nutrient-rich silt

Water
Only 50% of the population has access to improved water sources

Military budget
In 2012, just after the secession, South Sudan had the highest military spending in the world as a % of GDP

Secession
The country came in to existence after the 2011 referendum in which 98% voted in favour of secession from Sudan

Bandingilo National Park
2nd largest annual animal migration after the Serengeti

Gondokoro
Base for Sir Samuel Baker's expeditions, on one of which Lake Albert was discovered

Blue Nile conquest
In 2004, Pasquale Scaturro and Gordon Brown became the first people to complete the entire 5,230-km Blue Nile expedition. The whole 114-day journey was documented for IMAX theatres

Secession
After secession, Sudan lost as much land as the contiguous US would have lost if the states of California, Arizona, New Mexico, Texas, Louisiana and Arkansas had decided to leave the republic

Ilemi Triangle
Area disputed between Kenya and South Sudan (officially), and Ethiopia (unofficially)

Languages
Eritrea's constitution (1997), guarantees equality of all Eritrean languages and does not provide any list of suggested languages

Independence
Declared its independence from Ethiopia in1991 after a 30-year civil war and was internationally recognized as such after a referendum in 1993

Horn of Africa
Peninsula containing the countries of Ethiopia, Somalia, Eritrea and Djibouti

Population of 116.3 milllion people (Ethiopia: 99.4 million, Somalia: 10.8 million, Eritrea: 5.2 million and Djibouti: 0.9 million)

Modern humans probably originated in this region between 150,000 and 200,000 years ago

Only 220 species of mammals live in this region

Marathon
Ethiopia has 6 Olympic gold medals in the marathon and is the most successful nation in this category

Bab al Mandab Strait
Only 30 km wide. A dam built here would generate around 15 times the power of the largest US nuclear plant but at the same time it would destroy the ecosystem of the Red Sea

Gulf of Aden
Around 11% of the world' seaborne petroleum passe through these water

Around 21,000 ships sa through the gulf every yea

Kenenisa Bekele
The first athlete to win both 5,000 m and 10,000 m races at one World Championship (2009)

ERITREA

ASMARA

Exports
Coffee accounts for around 20% of Ethiopia's exports

Lake Assal & Afar Depression
It's one of the world's saltiest lakes (34.8% salinity), has the largest salt reserve and is the 2nd lowest point on Earth, with summer temperatures reaching up to 52°C

Djibouti gained independence from France in 1977

Forests
Ethiopia's forests are endangered. Severe droughts have been causing environmental problems since the 1980s. The civil war has also had an effect

"Lucy"
Australopithecus skeleton "Lucy", discovered in 1974, is 3.2 million years old

DJIBOUTI

DJIBOUTI

Camp Lemonnier
US military base, the only permanent one on this continent

Somali Civil War
On going since 1991

African Union
Established in 2001 in Ethiopia's capital, it integrates all African countries to participate in the global economy while addressing social and political problems

The triple juction point of the Indian, Arabian and African tectonic plates, the only one above sea level

SOMALILAND
Self-declared, not internationally recognized state

Ogaden territory

ADDIS ABABA
The highest capital in Africa (over 2,300 m)

ETHIOPIA

Ogaden territory
Area of military conflict between Ethiopia and Somalia since the 1970s. Somalis supporting the Ogaden National Liberation Front, are considered terrorists by the Ethiopian government

Only 0.01% of the population is Christian (1,200 people)

Kaffa
Homeland of coffee. The word itself comes from these southern Ethiopian lands

Ethiopian calendar
The Ethiopian calendar has 13 months. The first 12 months have 30 days, the 13th month has 5 days, but during a leap year it has 6

Landlocked country
Ethiopia is the most populous landlocked country on the planet (99 million people)

Livestock
One of the highest exporters of livestock in the world, mostly goats, sheep and camels

Ilemi Triangle
Administered by Kenya

First car in Africa
1st car to reach Ethiopia was Emperor Menelik II's car in 1907 – he was probably the 1st African to drive a car

Poetry
Known as "the nation of poets"

National parks
Kenya has 24 national parks (highest in Africa)

Lake Turkana

SOMALIA

African lion
Population in the wild:
1960s – 200,000
Present-day – 20,000

Described as a "failed state". Since 1991, Somalia has been destabilized and lacks a central government

INDIAN OCEAN

Swahili
National language in Kenya, Tanzania, Uganda and Dem. Rep. Congo

Olympics
With 31 Summer Olympics gold medals, Kenya is the most successful African country

Hunger
795 million people on Earth are starving

23% prevalence of undernourishment in the Sub-Saharan African region

73% of Somalis live on less than $2 per day

MOGADISHU
Before the civil war, Mogadishu was known as the "White pearl of the Indian Ocean"

KENYA

Lake Nakuru
"The greatest bird spectacle on Earth", the lake attracts thousands and sometimes millions of pink flamingos at nesting time

△ **Mount Kenya (5,199 m)**
2nd highest mountain in Africa. The country was named after it

NAIROBI

Nairobi National Park
Kenya's first national park, established in 1946

Piracy of the Somali coast
Triggered by toxic waste dumping and overfishing by foreign vessels

Somali fishermen began forming armed groups to protect their fishing grounds in the absence of an effective coastguard

They eventually turned to hijacking commercial vessels for ransom as an alternative and lucrative source of income

Between 2009 and 2011 the pirates extorted over $500 million, approximately $5 million per ship

Lake Victoria
World's largest freshwater tropical lake

The 2011 famine
Up to 260,000 people in East Africa starved to death

| 0 | 125 | 250 | 500 km |
| 0 | 75 | 150 | 300 miles |

Buganda Kingdom
Uganda takes its name from this kingdom, the largest of the traditional kingdoms within present-day Uganda

Landlocked country
With a population of 39 million, Uganda is the 2nd most populous landlocked country in the world (after Ethiopia)

Women in government
According to 2013 data, Rwanda had the highest proportion of women in government at 56%

Lake Edward

Rwandan Genocide
• lasted 100 days
• 2 million people slaughtered
• 0.5 million women brutally raped
• over 67% were infected with HIV as a result of its use as a weapon

Lake Kivu

Hunger
Around 1/3 of Burundi's population suffer from hunger

Deforestation
Much of Burundi's wildlife is threatened due to poaching and habitat loss, as only 5.4 % of land is protected. Elephants and gorillas are already extinct here

Source of the Nile
The location of the Nile's source is still disputed between Rwanda and Burundi

In Miocene times, Lake Tanganyika was the source of the Nile, making it 1,400 km longer. It was later blocked by the Virunga Mountains

Lake Tanganyika
2nd deepest freshwater lake in the world

White Nile
Albert Nile
Victoria Nile

Lake Albert

UGANDA

KAMPALA

Protected areas
Ugnda has over 60 protected areas with volcanoes, rainforests, lakes, rivers, savanna and abundant wildlife

National parks generate over $1.4 billion annually, around 10% of GDP

Queen Elizabeth National Park
Most visited national park in Uganda

Lake Victoria
World's 3rd largest freshwater lake with an area slightly larger than Croatia

RWANDA

KIGALI

BURUNDI

BUJUMBURA

People of Burundi have the world's lowest satisfaction with life

Urbanization
Burundi has the 2nd lowest rate of urbanization at 12.1%

Albinism
Tanzania has one of the world's highest rates of albinism. Albinos are often hunted for body parts by witch doctors to make potions

Serengeti National Park
World's greatest migration of up to 2 million animals

Ngorongoro Crater
World's largest (260 km²) unfilled caldera, formed after an explosion of a volcano up to 5,800 m high, 3 million years ago

Lake Tanganyika

Tree-climbing lions
There are only 2 populations of lions in the world that can climb trees, one in Queen Elizabeth National Park and the other in Lake Manyara National Park

Julius Nyerere
Tanzania's founding father and president (1962–85) was from the Zanaki group, the smallest ethnic group in the country

TANZANIA

Lake Manyara National Park
World's highest biomass density

DODOMA
The capital was moved here in 1996 to be in a more central location

Giraffe
National animal of Tanzania

Lake Nyasa
Also named Lake Malawi. It is a meromictic lake which means that it has layers of water that do not intermix

Tanzania and Malawi are in dispute over their border in Lake Nyasa

Lake Nyasa

Idi Amin
President of Uganda from 1971 to 1979. The film "The Last King of Scotland" is a historical drama based on events from Amin's time in power, for which Forest Whitaker won the Academy Award for Best Actor

Eastern black rhinoceros
Critically endangered. In 2010 they could only be found in Kenya (594 individuals) and Tanzania (80 individuals)

Animals of the Serengeti
Estimated population of the most popular species that live in the park:
• Tanzanian cheetah: 1,000
• African leopard: 1,000
• Masai lion: 3,000
• African bush elephant: 5,000
• Masai giraffe: 8,000
• Zebra: 250,000
• Gazelle: 500,000
• Serengeti wildebeest: 1.2 million

Kilimanjaro
Considered the highest summit that can be reached without special equipment. Highest mountain in Africa and the highest free-standing mountain in the world (5,892 m). Its snow caps have decreased by 80% since 1912

Tanzanite
Blue/violet mineral, endemic to the Mererani Hills. It was discovered in 1967 and is used as a substitute for sapphire

PEMBA

ZANZIBAR
Semi-autonomous part of Tanzania, known for cloves, nutmeg, cinnamon and black pepper production

DAR ES SALAAM
Former capital

MAFIA ISLAND
Developed in the 8th century as an important trade point between Asia and Africa. Together with Zanzibar and Pemba, known as the Spice Islands

National parks
Approximately 38% of Tanzania's land area is protected within 16 national parks

Tourism
Tourism contributes 1 million jobs & $4.5 billion annually (approximately 12% of GDP). Since 2004 its value has increased nearly threefold

	African Great Lakes	Great Lakes of North America
Surface area	158,800 km²	244,160 km²
Volume	31,000 km³	22,671 km³
Population	107 million	54 million
% of world's unfrozen fresh water	25%	18%

75	150	300 km
50	100	200 miles

Dictatorship

Equatorial Guinea's President Teodoro Obiang Nguema Mbasogo is currently the longest ruling dictator in the world (since 1979)

Vice-president
$
48-year-old playboy, Teodoro Nguema Obiang Mangue, appointed vice-president by his father, supposedly diverts millions of dollars of the country's revenues to his accounts

Poverty
Because of the dictatorship and uneven wealth distribution in Equatorial Guinea, around half of the population lacks access to safe drinking water and 10% of children die before reaching the age of 5

Chimpanzees and bonobo

The only two species of ape of the Pan genus and the closest relatives to humans. They are both endangered and can be found only in the Congo Basin

Neither of them can swim, so they evolved separately, divided by the formation of the Congo river, 2 million years ago. The bonobos live south of the river, and the chimps in the north

Goliath frog
This endangered species is the largest frog in the world, 33 cm long and 3 kg in weight. It inhabits Equatorial Guinea and parts of Cameroon

Nouabalé-Ndoki National Park
Inhabited by elephants, gorillas and chimpanzees. Due to its remoteness, some areas have never been explored. In 2006–7, over 125,000 western lowland gorillas were discovered here

MALABO

Spanish
Equatorial Guinea is the only African country with Spanish as its official language

EQUATORIAL GUINEA

Mbenga

Ethnic group of pygmies, that are thought to have been the earliest humans in Gabon, dating back to 7000 BC. Today, Mbenga tribes make up over 1/4 of the population

Chinese support
SÃO TOMÉ AND PRÍNCIPE
São Tomé and Príncipe's movement for independence was supported by China. It still provides aid

"Kuwait of Africa"
Small country, relatively rich in oil

Founded in 1849 by 52 freed African slaves

Rainforest

85% of Gabon's area is covered by rainforest, and 12% of it is protected

Chimpanzee fire
It's a type of fungus that glows in the dark, attracting insects which then spread its spores, helping the fungus to reproduce

Dwarf crocodile
It's the smallest crocodile species. Occurring in the caves of Gabon, they have unique orange skin

Lake Tele ★
Believed to be the habitat of a legendary creature, Mokélé-mbembé, similar to the Loch Ness Monster in Western culture

SÃO TOMÉ

LIBREVILLE

Ogooué

GABON

Ogooué river
All three African crocodile species live in this river

African rock python
Up to 5 m long, it's the Congolese forest's largest predator and the largest snake in Africa. Unlike most snakes, the mother is very protective of her nest and her young

Prince's Island
Discovered around 1470, it was named "Prince's Island" in honour of the Prince of Portugal

La Tropicale Amissa Bongo
2017 route (932 km) of the largest cycling race in Africa

CONGO

33 years as president
Denis Sassou Nguesso has been the president of Congo for 33 of the last 38 years

Loango National Park
Referred to as "Africa's last Eden", with white, sandy beaches, it is inhabited by hippos, elephants, gorillas, leopards and other species. It is the only West African lagoon protected as a national park

Congo-Ocean Railway
Built in the 1930s under French administration, it cost the lives of 17,000 workers who died from accidents and diseases. It also led to a civil war

Urban population
At 86%, Gabon has the highest percentage of urban population in Africa

BRAZZAVILLE

KINSHASA

The closest capitals
Separated only by the flow of the Congo river, Brazzaville and Kinshasa are two of the closest located capital cities in the world

POINTE-NOIRE

Congo

Toxic waste

Dictator, Teodoro Obiang, sold permits to Western corporations for the burial of millions of tons of toxic and radioactive waste here for $200 million per year

ANNOBÓN (Equ. Guinea)

Clouds

A cloudless day has never been recorded on Annobón

ATLANTIC OCEAN

Former Portuguese colonies

0	125	250	500 km
0	75	150	300 miles

White Nile

Economic potential
Oil, diamonds, timber, metals and other natural assets could make DRC one of the richest countries, but colonialism, slavery and corruption have turned it into one of the poorest

Cobalt
DRC produces around 60% of the world's cobalt. China is 2nd, but produces just 12% of the quantity

Rainforest vapour
Each hectare of rainforest produces 190,000 litres of water a year as vapour

Okapi
Okapi and giraffe are the only two surviving animals of the Giraffidae family. Okapi is much shorter and striped, like a zebra. It is endangered and lives only in the northern parts of DRC

Biodiversity
The most biodiverse African country with the greatest number of animal species

Poverty
77% of DRC's population live on less than $1.90 a day and 90% live on less than $3.10 a day

Congo

Mountain gorillas
The last 880 mountain gorillas live in this region

Tumba-Ngiri-Maindombe
At 66,000 km², it's the world's largest Wetland of International Importance – significant in terms of botany, ecology, hydrology, limnology and zoology

Late exploration
Most of the current territory of DRC was unexplored until the late 19th century

Virunga National Park
First national park in Africa (1925), currently managed by the Belgian Prince Emmanuel de Merode. The park is highly endangered due to poaching and natural resources exploitation

Congo river
World's deepest river, over 220 m deep. At 4,667 km long, it is 8th longest in the world and 2nd largest by discharge – 41,200 m³ of water per second, 5 times less than the Amazon

Renewable energy
99% of DRC's electricity is produced by hydroelectric plants alone

Boyoma Falls ★
World's largest waterfall by volume of annual flow rate. Consists of 7 cataracts which together have a fall of 61 m

Nyamuragira (3,058 m)
Africa's most active volcano, erupted over 40 times in the past 130 years

Internal conflicts
Since 1998, around 5 million people have died because of internal conflicts. Almost half were children

Nyiragongo (3,469 m)
One of five volcanoes in the world with a visible and persistent lava lake

Lake Kivu
One of just three exploding lakes on the planet. The other two are in Cameroon

Lake Mai-Ndombe
During the rainy season, this lake can triple in size

DEMOCRATIC REPUBLIC OF THE CONGO

Kifuka ★
This village receives the highest number (158/km²) of annual lightning strikes in the world

UN peacekeeping
The largest UN peacekeeping mission has been stationed in DRC since 2010 (around 19,000 military personnel)

Former Zaire
During the reign of dictator Mobutu Sese Seko, the country was named Zaire (1971–97)

KINSHASA
With 12.6 million people, it is the 2nd largest French-speaking city in the world

Colonialism
DRC was a Belgian colony from the 19th to 20th centuries. Under the regime of King Leopold II, more than 10 million Congolese lost their lives because of the colonists' barbaric rules

Francophone country
With over 77 million citizens, it's the most populous French-speaking country in the world

Size
2nd largest country in Africa and 11th in the world. It is similar in size to Greenland

Authenticité
State ideology during the Mobutu regime (1965–97) aimed at clearing out all traces of Western culture. Citizens had to change their names to traditional African ones

The most dangerous place to be a woman
Nearly 40% of women in the eastern Congo experience sexual violence at least once in their lives

Rape
In 2008, the UN officially declared rape a weapon of war

Laurent-Désiré Kabila
President of DRC who overthrew Mobutu in 1997 with the help of militia responsible for the Rwandan genocide. He was shot in 2001 by one of his boy soldiers. 10,000 children served in his army

Uranium
Uranium for the nuclear bombs dropped on Japan during WWII came from a mine in southeast Congo

Chinese trade partner
China is the country's main trading partner, receiving around 50% of its exports

Congo

Africa's rainforests
The Congo Basin constitutes around 13% of the African continent and is 2nd in size in the world, to the Amazon. It has the lowest deforestation rate of all rainforests

"Heart of Darkness"
Famous novella written in 1899 by Joseph Conrad about a voyage up the Congo river. It tells a story revolving around Mr Kurtz, a mysterious ivory trader

Francis Ford Coppola directed "Apocalypse Now" (1979), a loose adaptation of this novel

Zambezi

125	250	500 km

| 75 | 150 | 300 miles |

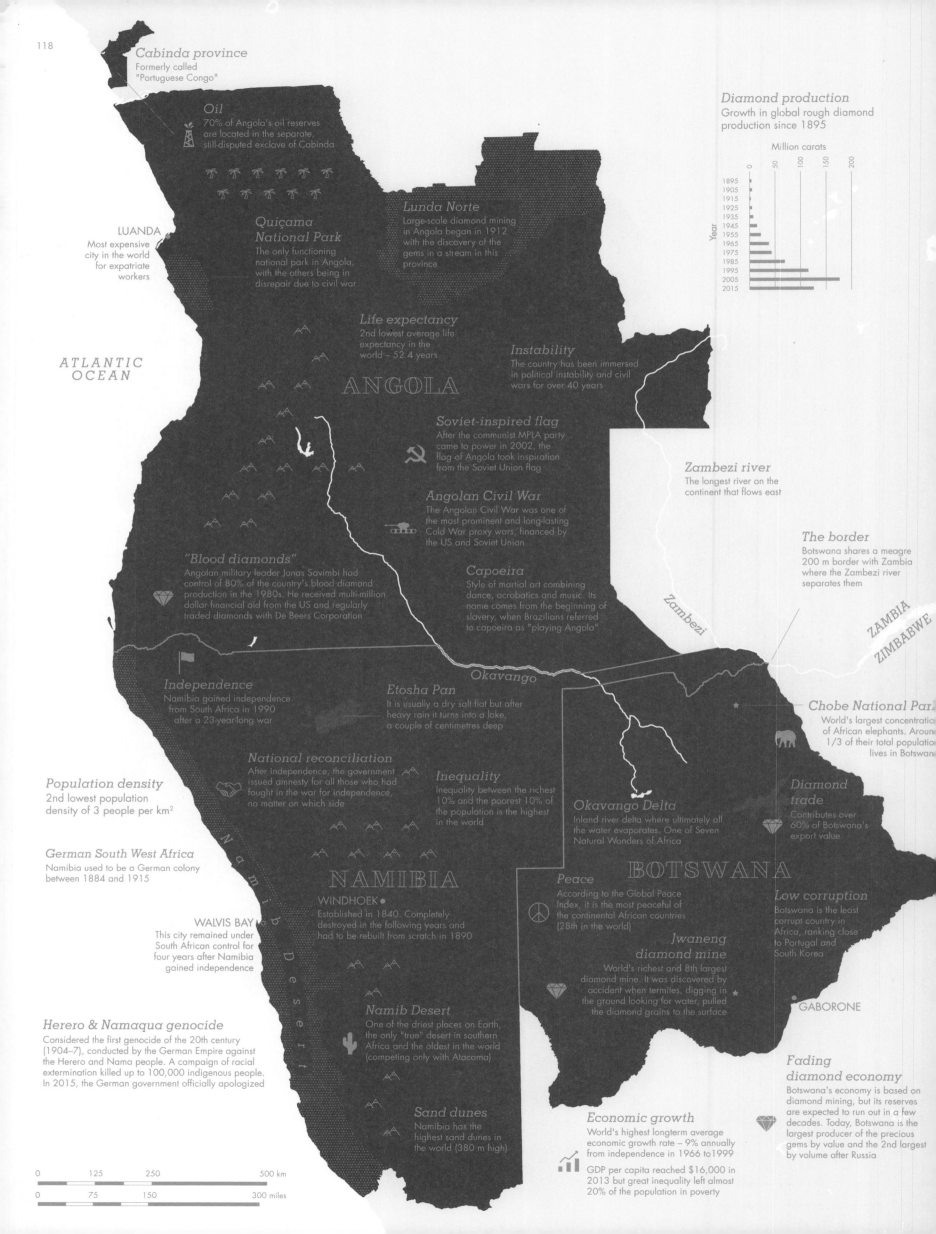

Cabinda province
Formerly called
"Portuguese Congo"

Oil
70% of Angola's oil reserves
are located in the separate,
still-disputed exclave of Cabinda

Lunda Norte
Large-scale diamond mining
in Angola began in 1912
with the discovery of the
gems in a stream in this
province

LUANDA
Most expensive
city in the world
for expatriate
workers

*Quiçama
National Park*
The only functioning
national park in Angola,
with the others being in
disrepair due to civil war

Diamond production
Growth in global rough diamond
production since 1895

Million carats

Life expectancy
2nd lowest average life
expectancy in the
world – 52.4 years

Instability
The country has been immersed
in political instability and civil
wars for over 40 years

ATLANTIC
OCEAN

ANGOLA

Soviet-inspired flag
After the communist MPLA party
came to power in 2002, the
flag of Angola took inspiration
from the Soviet Union flag

Zambezi river
The longest river on the
continent that flows east

Angolan Civil War
The Angolan Civil War was one of
the most prominent and long-lasting
Cold War proxy wars, financed by
the US and Soviet Union

The border
Botswana shares a meagre
200 m border with Zambia
where the Zambezi river
separates them

"Blood diamonds"
Angolan military leader Jonas Savimbi had
control of 80% of the country's blood diamond
production in the 1980s. He received multi-million
dollar financial aid from the US and regularly
traded diamonds with De Beers Corporation

Capoeira
Style of martial art combining
dance, acrobatics and music. Its
name comes from the beginning of
slavery, when Brazilians referred
to capoeira as "playing Angola"

Zambezi

ZAMBIA

ZIMBABWE

Okavango

Independence
Namibia gained independence
from South Africa in 1990
after a 23-year-long war

Etosha Pan
It is usually a dry salt flat but after
heavy rain it turns into a lake,
a couple of centimetres deep

Chobe National Park
World's largest concentration
of African elephants. Around
1/3 of their total population
lives in Botswana

Population density
2nd lowest population
density of 3 people per km²

National reconciliation
After independence, the government
issued amnesty for all those who had
fought in the war for independence,
no matter on which side

Inequality
Inequality between the richest
10% and the poorest 10% of
the population is the highest
in the world

Okavango Delta
Inland river delta where ultimately all
the water evaporates. One of Seven
Natural Wonders of Africa

*Diamond
trade*
Contributes over
60% of Botswana's
export value

German South West Africa
Namibia used to be a German colony
between 1884 and 1915

NAMIBIA

BOTSWANA

WINDHOEK

Peace
According to the Global Peace
Index, it is the most peaceful of
the continental African countries
(28th in the world)

Low corruption
Botswana is the least
corrupt country in
Africa, ranking close
to Portugal and
South Korea

WALVIS BAY
This city remained under
South African control for
four years after Namibia
gained independence

Established in 1840. Completely
destroyed in the following years and
had to be rebuilt from scratch in 1890

*Jwaneng
diamond mine*
World's richest and 8th largest
diamond mine. It was discovered by
accident when termites, digging in
the ground looking for water, pulled
the diamond grains to the surface

GABORONE

Herero & Namaqua genocide
Considered the first genocide of the 20th century
(1904–7), conducted by the German Empire against
the Herero and Nama people. A campaign of racial
extermination killed up to 100,000 indigenous people.
In 2015, the German government officially apologized

Namib Desert
One of the driest places on Earth,
the only "true" desert in southern
Africa and the oldest in the world
(competing only with Atacama)

*Fading
diamond economy*
Botswana's economy is based on
diamond mining, but its reserves
are expected to run out in a few
decades. Today, Botswana is the
largest producer of the precious
gems by value and the 2nd largest
by volume after Russia

Sand dunes
Namibia has the
highest sand dunes in
the world (380 m high)

Economic growth
World's highest longterm average
economic growth rate – 9% annually
from independence in 1966 to 1999

GDP per capita reached $16,000 in
2013 but great inequality left almost
20% of the population in poverty

0	125	250	500 km
0	75	150	300 miles

"Big five game"
Includes the 5 herding animals of Africa, all of which inhabit both Zambia and Zimbabwe: African lion, African elephant, Cape buffalo, African leopard, rhinoceros

The term was coined during the colonial period by big-game hunters

"Z"
Zambia and Zimbabwe are the only two countries whose names start with the letter "Z"

Lake Tanganyika

Lake Mweru

Congo

African fish eagle
The national bird of Zimbabwe and Zambia, it is included on their flags. It preys mainly on fish and mates for life

Mbala
The last surrender of German soldiers to the British during WWI was at Abercorn, now called Mbala

Copper
Used to be the 2nd largest copper producer. Today it's 8th

Lake Bangweulu
Crucial to the biodiversity of northern Zambia and to a significantly larger region for birdlife. It is facing environmental issues. Its average depth is only 4 m

Bangweulu Swamps
One of the largest wetland systems on the planet, extending over 15,000 km²

Zambezi

Congo river's source
The source of the world's 2nd largest river (by discharge), after the Amazon

Peace
Unlike most of its neighbours, Zambia avoided wars during the post-colonial period in Africa

Population growth
By 2100, the population of Zambia is predicted to grow more than 5 times. It's 16.2 million today

Economic reforms
In 2010, Zambia was praised by the World Bank for its fast pace of economic reforms

Kafue National Park
The largest in Zambia and one of the most famous in the world. It has a similar size to Djibouti

ZAMBIA

One of the fastest developing cities in southern Africa
• LUSAKA

Zambezi rainforest
The only forest that receives water 24/7 from the enormous splash of Victoria Falls

Zambezi

Operation Noah
The first ever large-scale relocation of wild animals (1958–64) during the construction of Kariba Dam. Over 6,000 animals were relocated and saved

Tribe's relocation
Due to construction of Kariba Dam, the Tonga tribe (57,000 people) was relocated and still suffers the effects

Zambezi

Employment of last resort
Today, nearly 95% of the population has to rely on informal occupations, such as subsistence farming, in order to survive

Lake Kariba
World's largest artificial lake by volume, slightly bigger than Vänern, the largest natural lake in the EU

Lake Kariba

Former Rhodesia
The country was named Rhodesia until 1979, after Cecil Rhodes, owner of the British South Africa Company, a mining enterprise that bought these lands at the end of the 19th century

HARARE
•

Moonbow
Beautiful optic phenomenon visible on the waters of the Victoria Falls, formed by the reflection of moonlight

Victoria Falls
One of The Seven Natural Wonders of the World. Largest waterfall by sheet of falling water

ZIMBABWE

Dysfunctional economy
Robert Mugabe's "legislation", including confiscation of private assests in recent decades, has resulted in the destruction of most of the independent, organized forms of economic activity in the country

Diamond production
Out of the top 10 diamond producing countries by volume, 7 are African. Together, these 10 countries produce around 99% of the world's total (by volume)

Dictatorship
Zimbabwe has been ruled by Robert Mugabe since 1980 when his party won the elections that ended white minority rule

De Beers Corp.
Established by Cecil Rhodes. The first to manipulate consumer demand by advertising diamonds as a symbol of eternal love

2015 diamond production (million carats)
1. Russia 37.9
2. Botswana 23.2
3. DRC 15.7
4. Australia 11.7
5. Canada 10.6
6. Zimbabwe 10.4
7. Angola 9.4
8. South Africa 8.1
9. Namibia 1.7
10. Sierra Leone 0.6

Oldest leader
President of Zimbabwe, Robert Mugabe, is the world's oldest leader currently in power. He is 93

"A diamond is forever"
This De Beers slogan has been named the best advertising slogan of the 20th century

0   100   200   400 km

0   50   100   200 miles

"The Calendar Lake"
Lake Nyasa is sometimes called the Calendar Lake as it's 365 miles long

Lake Nyasa
Home to more species of fish than any other lake on the planet. About 80% of all tropical fish sold for freshwater aquariums can be found here

At 706 m deep, it's the 2nd deepest lake in Africa and 6th deepest on the planet

"Smoke on the water"
The lake is inhabited by gigantic swarms of harmless flies, which, from a distance, resemble dark clouds floating above the water surface

"Lake of Stars"
The first nickname was coined by Scottish explorer David Livingstone, as he observed the lantern lights of the fishermens' boats at night

1st British naval victory in WWI
18 days after the start of the war, British steamship SS *Gwendolen* destroyed the only German gunboat on Lake Nyasa, the *Hermann von Wissmann*, with a single shot

Nyau
A secret society of Malawi's Chewa tribe, whose mission was to guard sacred knowledge regarding cosmology, religious beliefs and forms of ritual dance, Gule Wamkulu

Little is known about the history of Nyau as it was based on oral tradition and breaking the society's secrets was punished by death

MALAWI

LILONGWE

MOZAMBIQUE
The name appears to originate from Mussa-bin-Mbiki, the son of an Arab sultan who ruled these lands

Portugal
Mozambique was a Portuguese colony from 1505 to 1975

Nyasaland
Former name of the country which derives from the Nyasa people who inhabit the region

Lake Chilwa
It has no outlets, so its water level is greatly dependent on seasonal rains and evaporation. During the dry season in 1968, the lake disappeared completely

"Google Forest"
Primeval forest, undiscovered until 2005, when scientists started exploring its unique wildlife with use of Google's satellite imagery

Black mamba
The longest venomous snake in Africa and the world's fastest. It's highly aggressive and its bite can kill an adult elephant. The inside of its mouth is black

Sea turtles
Mozambique's coastline is home to five of seven endangered species of sea turtles

Mozambique Channel eddies
These large vortex currents, up to 300 km in diameter, can reach speeds of up to 1 m/s. The exact reason for their formation is still debated

Zambezi

Zambezi

Mussiro Mask
White paste, obtained from mussiro wood, which women in Mozambique use to cover their faces. Once it used to be a symbol of virginity, today it remains popular for beautifying the skin

Vasco da Gama
The first European to explore these lands in 1498

BASSAS DA INDIA (Fr.)
This atoll is the edge of a remnant volcano rising 3,000 metres from the bottom of Mozambique Channel. It's a famous spot for diving among sharks and shipwrecks

Splitting Africa
In 10 million years from now, the Somali tectonic plate is expected to split Mozambique from the African plate, just as Madagascar did 135 million years ago

EUROPA ISLAND (Fr.)
One of the first episodes of "The Undersea World of Jacques Cousteau" was shot here in 1968, documenting the breeding of green sea turtles. It was a pioneering achievement for both marine biology and underwater filming

Independence
After Mozambique gained independence from Portugal in 1975, the Portuguese were forced to leave the country within 24 hours

Praia do Tofo
One of the prime spots in Africa for divers to experience an encounter with whale sharks or manta rays

Around 135 million years ago, Madagascar split from the Africa-South America landmass

Maputo station
The picturesque railway station in Maputo is considered one of the most beautiful in the world

Presidential plane crash
In 1986, Mozambican president Samora Machel and 33 others died in a plane crash. Only 10 people survived

MAPUTO

0	125	250	500 km

0	75	150	300 miles

ALDABRA
(Seychelles)
2nd largest coral
atoll in the world

Brookesia micra
The smallest chameleon and one
of the smallest reptiles on the
planet, roughly 3 cm long. It was
discovered in the mid-2000s and
can be found only on the small
rocky island of Nosy Hara

MORONI COMOROS

Essential oils
The largest producer
of ylang-ylang, the
main ingredient of
many perfumes

Territorial claim
Comoros claims the
island of Mayotte

MAYOTTE (Fr.)
Voted against becoming independent
from France in 1974, 1976 and 2009

Due to the island's prosperity, it
is the region's major destination
for illegal immigrants

SEYCHELLES

115 islands
41 are pre-Cambrian
rock, 74 are coral

 VICTORIA

Population
Smallest population of
any sovereign African
country – 96,000 citizens

Olivier Levasseur
Pirate, known for hiding one of the
biggest treasures in history here,
estimated at over £1 billion

Climate
Average year-round
temperature of 27°C,
with 80% humidity

Around 88
million years ago,
Madagascar split
from India

Lemur
It's a clade of primates
which includes nearly 100
different species, all of them
endemic to Madagascar

Because of the island's isolation,
and seasonal and diverse climate,
no other place on Earth has
hosted a similar variety of one
primate

Arrival
The first people who settled in
Madagascar arrived from Borneo
in canoes around 350 BC

Elephant bird
3 m tall and weighing
500 kg, it was the largest
bird on the planet. It
became extinct by the
17th or 18th century

Archaeoindris
With a body mass of 200 kg, it was
the largest species of lemur, the size
of a gorilla. Its extinction coincided
with the arrival of the first humans to
the island around 350 BC

MADAGASCAR
4th largest island in the world

ANTANANARIVO●

"Gondwana"
The name of the ancient
supercontinent, which once
included Madagascar,
Antarctica, India, Africa,
South America and Australia

I N D I A N
O C E A N

M O Z A M B I Q U E   C H A N N E L

Avenue of
★ the Baobabs
Beautiful and famous
dirt road with many
prominent baobab
trees lining it

Vanilla
2nd most expensive
spice after saffron.
Madagascar is the 2nd
largest vanilla producer
after Indonesia

Madame Berthe's
mouse lemur
It weighs just 30 grams,
making it the smallest
lemur and the smallest
primate on the planet

Madagascar's currency
① 1 ariary = 5 iraimbilanja. The
Malagasy ariary is one of only two
currencies in which units are not
based on the power of 10. The
other is the Mauritanian ouguiya

Endemic species
90% of all the island's animals and
plants are endemic. For millions of
years they evolved independently,
separated from Africa and India,
making the island's flora and fauna
absolutely unique

Dodo
Large flightless bird, 1m tall; was
found only on Mauritius, where it
remains a national symbol

The bird had no fear of humans
and was hunted to extinction by
sailors in the late 17th century

 Hinduism is the religion
of approximately 50%
of Mauritians

MAURITIUS
1 million tourists annually

Réunion is the
outermost region of
the European Union

PORT LOUIS

Peace
According to the Global
Peace Index, Mauritius is
the most peaceful country
in Sub-Saharan Africa

RÉUNION (Fr.)
Originally named
Bourbon by the French
who discovered the
island

△ Piton de la Fournaise (2,632 m)
One of the most active volcanoes in the world.
When eruptions occur, they occasionally
generate rivers of lava flowing into the ocean,
resulting in spectacular sights

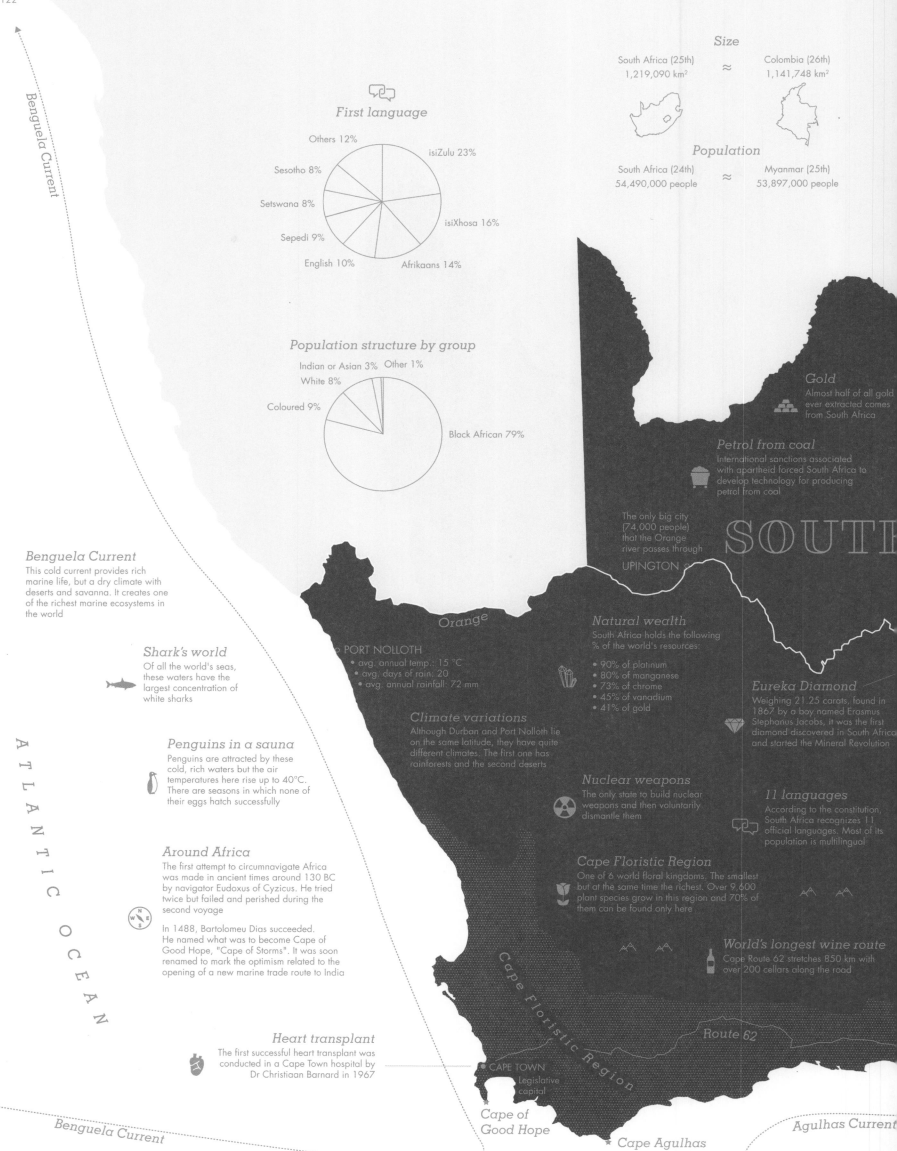

Size

South Africa (25th)
1,219,090 km²

≈

Colombia (26th)
1,141,748 km²

Population

South Africa (24th)
54,490,000 people

≈

Myanmar (25th)
53,897,000 people

First language

Others 12%
Sesotho 8%
Setswana 8%
Sepedi 9%
English 10%
isiZulu 23%
isiXhosa 16%
Afrikaans 14%

Population structure by group

Indian or Asian 3% Other 1%
White 8%
Coloured 9%
Black African 79%

Gold
Almost half of all gold ever extracted comes from South Africa

Petrol from coal
International sanctions associated with apartheid forced South Africa to develop technology for producing petrol from coal

The only big city (74,000 people) that the Orange river passes through
UPINGTON

SOUTI

Benguela Current
This cold current provides rich marine life, but a dry climate with deserts and savanna. It creates one of the richest marine ecosystems in the world

Orange

PORT NOLLOTH
• avg. annual temp.: 15 °C
• avg. days of rain: 20
• avg. annual rainfall: 72 mm

Natural wealth
South Africa holds the following % of the world's resources:

• 90% of platinum
• 80% of manganese
• 73% of chrome
• 45% of vanadium
• 41% of gold

Shark's world
Of all the world's seas, these waters have the largest concentration of white sharks

Climate variations
Although Durban and Port Nolloth lie on the same latitude, they have quite different climates. The first one has rainforests and the second deserts

Eureka Diamond
Weighing 21.25 carats, found in 1867 by a boy named Erasmus Stephanus Jacobs, it was the first diamond discovered in South Africa and started the Mineral Revolution

Penguins in a sauna
Penguins are attracted by these cold, rich waters but the air temperatures here rise up to 40°C. There are seasons in which none of their eggs hatch successfully

Nuclear weapons
The only state to build nuclear weapons and then voluntarily dismantle them

11 languages
According to the constitution, South Africa recognizes 11 official languages. Most of its population is multilingual

Around Africa
The first attempt to circumnavigate Africa was made in ancient times around 130 BC by navigator Eudoxus of Cyzicus. He tried twice but failed and perished during the second voyage

In 1488, Bartolomeu Dias succeeded. He named what was to become Cape of Good Hope, "Cape of Storms". It was soon renamed to mark the optimism related to the opening of a new marine trade route to India

Cape Floristic Region
One of 6 world floral kingdoms. The smallest but at the same time the richest. Over 9,600 plant species grow in this region and 70% of them can be found only here

World's longest wine route
Cape Route 62 stretches 850 km with over 200 cellars along the road

A T L A N T I C   O C E A N

Cape Floristic Region

Route 62

Heart transplant
The first successful heart transplant was conducted in a Cape Town hospital by Dr Christiaan Barnard in 1967

CAPE TOWN
Legislative capital

Cape of Good Hope

Benguela Current

Agulhas Current

Cape Agulhas
Africa's southernmost point

Mass relocation

In the period 1960 – 83, over 3.5 million non-white citizens were forced to relocate into segregated ghettos

Kruger National Park

First South African national park (1926). The Tsonga people were forcibly removed from an area they dominated. Later, in 1969, the Makuleke people were also forced off the northern part of the park but regained their territory after claiming it in 1996

Kruger National Park

SWAZILAND

Longest living monarch

The longest reign of any monarch in recorded history was 82 years by King Sobhuza II of Swaziland

HIV

In 2006, the prevalence of HIV/AIDS reached 25% in Swaziland and over 24% in Botswana – the two highest rates in the world

Apartheid

System of racial segregation and discrimination from 1948 to 1991. The first election, free for all, took place in 1994 and Nelson Mandela became the country's first black president

Vredefort crater

World's largest verified impact crater, around 300 km wide

Cullinan Diamond

The world's largest gem-quality diamond ever, found here in 1905

Cut into 9 large and 96 smaller gems, it forms part of the Crown Jewels of the UK. Includes the largest colourless cut diamond "The Great Star of Africa"

PRETORIA
Executive capital

JOHANNESBURG

SOWETO

Two Nobel Peace Prize winners lived on the same street in Soweto – Nelson Mandela and Archbishop Desmond Tutu

AFRICA

MBABANE
LOBAMBA
SWAZILAND

Aghulas Current

Busiest airport

O.R. Tambo airport is the busiest on the continent – 19 million passengers in 2014

Tugela Falls

2nd tallest waterfall in the world (948 m). Some argue that it actually is taller than the Angel Falls (979 m)

Enclave

Lesotho is one of 17 states that border only one country but only one of three (along with Vatican City and San Marino) that at the same time do not border any sea or ocean

Katse Dam

2nd largest dam in Africa

BLOEMFONTEIN
Judicial capital

MASERU
First established as a police camp

LESOTHO

Orange river

The longest river in the country. It was named in honour of William V of Orange, Captain-General of the Dutch States Army in the 18th century

Orange

Drakensberg

Mtentu

DURBAN
• avg. annual temp.: 21°C
• avg. days of rain: 160
• avg. annual rainfall: 1,050 mm

Mysterious pilgrimage

Giant trevally fish (up to 80 kg), travel many kilometres upstream into rivers like the Mtentu. At their destination they begin a rare spectacle – as if they were dancing, they circle around melancholically for weeks, then return to the sea. They neither hunt nor reproduce and the nature of their journey remains unknown

South

South Africa is the southernmost state of the Old World's mainland and Eastern Hemisphere

Karoo

Most of the country is covered by a plateau composed almost entirely of the Karoo rock that started forming around 320 million years ago

How many capitals?

The only country with three capitals, where each hosts a different branch of government activity. None of the three is superior to the others

Great sardine run

Rare phenomenon of sardines grouping into billions and becoming visible from space

They migrate from the cold waters of the Atlantic to the warmer Indian Ocean to reproduce, but only when the latter's temperature falls below 21 °C

LESOTHO

High elevation

The country with the highest lowest point in the world (1,400 m). No other country lies completely above 1,000 m

Hydropower

100% of the electricity comes from hydropower. Katse Dam is the highest (2,050 m) and the second largest dam in Africa

INDIAN OCEAN

Antarctica

Approximately 3,800 km distance from coast to coast

0	50	100	200 km

0	30	60	120 miles

The 1st solar nation
With a population of 1,250,
Tokelau is the world's first 100%
solar-powered nation

OCEANIA

OCEANIA

Area
With an area of 8,844,516 km², it is the smallest continent, slightly smaller than Europe

Population
Besides Antarctica, it is the least populated continent with a population comparable to Poland. 2/3 of people live in Australia

Oceania (0.54%)
39,901,000 people

$
Wealth
The total GDP of Oceania countries is $1.47 trillion – comparable to South Korea's

Countries
Oceania includes 13 sovereign states and 16 non-sovereign territories

Definition
According to the geopolitical division used by the UN, Oceania includes Australia and the Pacific nations, excluding the Malay Archipelago apart from Papua New Guinea

Nuclear testing
Oceania was the site of many nuclear tests due to its sparse population. UK, France and the US carried out bomb detonations here, often with a negative impact on peoples' lives

Bougainville conflict
Bougainville Province has made several attempts to secede from Papua New Guinea. Due to the location of a copper mine crucial to Papua New Guinea's economy, all attempts have failed

Name
The name Oceania comes from the French "Océanie", popularized in 1812 by the geographer Conrad Malte-Brun who also imported camels to Australia

"Larger Australia"
Extension of Australia's landmass at its maximum around 20,000 years ago when the sea level was much lower

INDIAN OCEAN

Tourism
In 2016, international tourists spent $38.8 billion in Australia

Boomerang
Weapon used for hunting, developed by indigenous Australians

Most dangerous ant
Bulldog ants, native to Australia, are aggressive, do not fear humans and their stings can kill an adult

Wooliest sheep
In 2015, a record-breaking 40 kg of wool was shorn from an Australian sheep

Australian rules football
A combination of football and rugby. 36 players (18 per team) can tackle, kick the ball or punch it but can't pass it by throwing

Kati Thanda-Lake Eyre
Lowest point in Oceania (-16 m)

BRISBANE
Population: 2.3 million

Darling

Murray

SYDNEY
Population: 4.6 million

MELBOURNE
Population: 4.3 million

Most lethal animal
The jellyfish, Chironex fleckeri, is considered one of the world's most lethal animals. It lives off the coasts of Australia, New Guinea, the Philippines and Vietnam and has enough venom to kill 60 people

Olympics
Australia is the only country in Oceania to host the Summer Olympics, in Melbourne in 1956 and in Sydney in 2000

0	500	1,000		2,000 km

0	300	600	1,200 miles

Japanese Empire
During WWII, the aim of the Japanese was to gain control over the islands of the Pacific but its strategic defeats at the battles of Midway and Coral Sea hindered this objective

Agriculture
The main economic activity for Pacific countries, except for Australia and New Zealand. The primary crops are: copra, coconuts, timber, palm oil, cocoa, beef, dairy, bananas, kava and sugarcane

Trade
Smaller Pacific nations depend on trade with Australia and New Zealand. Other large export markets are USA, Japan, China, South Korea and India

Giant clam
The largest (up to 120 cm long) and one of the most endangered clam species. It can live up to 100 years

Peace
Since World War II, the region of Oceania has generally been peaceful

Nauru
This island country has the highest population density (476 people/km²) in Oceania, followed by Tuvalu (397 people/km²)

Christianity
Oceania's predominant religion (around 73% of the population)

"Samoan time travel"
In 2011, Samoans skipped an entire day – 30 December. The country wanted to change its time zone, historically related to the US, to a new one, that would bring them closer economically to Oceania and Asia

"Living fossil"
Nautilus, a marine mollusc, is the 4th oldest living species on the planet. It lives in tropical waters deeper than 600 m

PACIFIC OCEAN

Zealandia
Partly submerged landmass, considered by some to be a separate, 8th continent

International Date Line
Imaginary line which defines the boundary between days in the calendar. Theoretically, if you cross it from west to east, you will "gain" one day

Football in Oceania
Oceania is the only football confederation that has no automatic qualification to the World Cup. It has been represented there five times – New Zealand in 1982 and 2010 and Australia in 1974, 2006, 2010 and 2014

15 minutes
Chatham Island (GMT +12:45) and Central Western Australia (GMT +08:45) are two of only three places that recognize 15-minute time zone offsets

Whale shark
Largest known living fish species, up to 12.65 m long. It can be found in waters warmer than 22 °C

International Date Line

TIMOR SEA

Java Trench
★ Deepest point in the Indian Ocean (-7,125 m)

Trade
Japan is the largest importer of Australian goods

Kakadu National Park
Australia's largest national park is equal to the size of Israel and nearly half the size of Switzerland

Moby Dick
The fictional route of the ship "Pequod" chasing the white whale, based on the novel "Moby Dick". Published in 1851, it was a commercial failure but it earned its reputation in the 20th century

Horizontal Falls
The only place on earth where waterfalls run horizontally

Human development
Australia has the 2nd highest HDI (human development index) in the world, after Norway

Around a quarter of the population is foreign-born and another fifth had at least one parent born overseas. Every 3 minutes Australia gains another migrant

Immigration
Over 160,000 convicts were brought to Australian penal colonies by the British from 1788 to 1868. Around 20% of modern Australians are their descendants

Wolfe Creek Crater
★ 2nd largest visible meteorite crater in the world

Discovery
First documented encounter of Australia was that of Dutch navigator Willem Janszoon in 1606, however, some advanced the theory that the Portuguese were the first Europeans to reach it in the 1520s

Black swan
Characteristic bird of Western Australia – it is represented on both the state's flag and emblem

Rabbit-Proof Fence
In 1859, 24 rabbits brought from Great Britain were released for hunting. Due to the warm climate and lack of natural predators they spread rapidly

Three fences were built, stretching 3,256 km, to save the pastoral areas of Western Australia

Numerous measures were taken to control the population including poison, viruses and diseases but the rabbits became resistant to them

State of Western Australia
2nd biggest country subdivision in the world, similar to the size of Kazakhstan

Lake Disappointment
Found by Frank Hann in 1897 who hoped to find fresh water here

The Queen
Elizabeth II is formally the Queen of Australia

Uluru (Ayers Rock)
Famous rock, sacred to indigenous people. It was created over 600 million years ago. It originally sat at the bottom of a sea

AUS

Shark Bay
World Heritage site, home to many globally threatened species including dolphins, sharks, birds and cyanobacteria

Lakes
Most of Australia's lakes are salty and ephemeral (periodic)

Great Victoria Desert
The largest among 10 Australian deserts. Approximately 35% of the continent area is impacted by desertification

Highway 1
Encircling 25,000 km of the whole continent, it's the longest, continuous highway in the world

Great white shark
The highest number of recorded shark attacks on humans – overall there are more than 70 shark attacks reported worldwide each year – most fatal attacks occur off the coast of Australia

Indian Pacific Railway

Highway 1

PERTH
The most isolated, large city – closest neighbouring metropolitan area is over 2,700 km away

Australia has the 3rd lowest population density after Namibia and Mongolia. It is mostly concentrated along the coasts

Indian Pacific Railway
One of the most scenic railways and one of the greatest investments and engineering achievements in the history of Australia

The route includes the world's longest straight stretch of railway track (478 km across the Nullarbor Plain)

Lake Hillier
Saline lake on a small island, notable for its pink waters. The only living organism there is a microorganism causing the salt content in the lake to create a red dye, like the colour of bubble gum

Highway 1

Similar size to Brazil

Brazil
8,514,879 km²

Australia
7,692,024 km²

2,500 km
Distance from Australia to Antarctica

INDIAN OCEAN

Moby Dick

Rabbit-Proof Fence

Highway 1

0	150	300		600 km

0	100	200		400 miles

Gulf of Carpentaria

The largest country in the world without land borders

Daintree Rainforest
Probably the oldest rainforest in the world (up to 180 million years old)

Great Barrier Reef
World's largest coral reef system, composed of over 2,900 individual reefs and 900 islands, stretching for more than 2,300 km over an area of 344,000 km²

CORAL SEA
Oil exploration in these waters was terminated in 1975

Koala
It doesn't need to drink, because it can obtain all of the moisture it needs by eating leaves

Didgeridoo
The oldest known wind instrument, made of eucalyptus wood, hollowed from within by termites

Larger than Finland, Vietnam or Poland

Approximately 1,500 fish species (10% of the world's total) can be found within it

Sediment, nutrients, agricultural pesticide pollution and warmer ocean temperatures put stress on coral and lead to coral bleaching

Indigenous Australians

Native people of Australia who came to the continent tens of thousands of years ago

When Europeans started migrating here, they were marginalized and forced to live in the deserts

They constitute just 2% of Australia's present day population

Highway 1

Great Barrier Reef

Great Dividing Range

PACIFIC OCEAN

ALIA

The largest country that lies completely in the Southern Hemisphere

6th largest country by area and 52nd by population

Independence
Australia gained independence from Great Britain in 1901

Great Dividing Range
3rd longest mountain range in the world which provides the entire fresh water supply for Eastern Australia

Kati Thanda-Lake Eyre
When it occasionally fills with water it's the largest lake in Australia. It's also the lowest natural point on the continent

Dingo Fence
Finished in 1885, it keeps dingoes out of the relatively fertile southeast part of the continent and protects the sheep flocks. Stretching 5,614 km, it's the world's longest fence

Little Penguin
The smallest species of penguin. It grows to an average of 33 cm tall, and lives in Southern Australia and New Zealand

Dingo Fence

Darling

Lake Torrens
Salt flat, last filled with water 150 years ago

Burning Mountain
Has an underground natural coal fire. The scientific estimate is that the fire has burned for 6,000 years. It is the oldest known coal fire

Platypus
One of the few mammals in the world which lays eggs. It is venomous and endemic only to Australia and Tasmania

Indian Pacific Railway

Less than 1% of the country is covered with water due to the absence of glacial and tectonic activity in the past

1st world surfing championships were held at Manly in 1964 and won by an Australian, Bernard "Midget" Farrelly

SYDNEY

Murray

Melbourne was almost called Batmania because its northern part was explored by John Batman

Murray

CANBERRA

The last state to decriminalise homosexuality; until 1997, the punishment was up to 25 years in jail

The longest tram network in the world with over 250 km of tram rails

Mt Kosciuszko (2,229 m)

MELBOURNE

Tasmanian Tiger
Thought to be extinct for 40 years, the Tasmanian Tiger was the last extant member of Thylacinidae. Specimens of this family have been found in fossil records dating back to early Miocene

TASMANIA

TASMAN SEA
Called "The Ditch" by both Australians and New Zealanders

Around 3,800 km
to French Polynesia

Circumnavigation

During his 1st voyage (1768–71), Captain James Cook completed the first circumnavigation of New Zealand. He disproved the hypothesis that New Zealand was connected to a larger southern landmass and mapped its coastline with great precision considering the tools available in the18th century

Tuatara

Tuatara are the oldest living species of reptile. They can be traced back 190 million years to the Mesozoic era. They have a life expectancy of about 60 years but can live to be over 100 years

NORTH ISLAND
(TE IKA-A-MĀUI)

Boiling spring

Frying Pan Lake is one of the world's largest hot springs, maintaining a water temperature of 50–60°C. At the centre, where the lake is actively boiling, the temperature can't be accurately measured

AUCKLAND
The city has one of the highest boat ownerships per capita in the world

Bay of Plenty

Waitomo Caves

They are home to a unique species of glowworm producing an amazing light effect – vast clusters of these insects resemble starry skies

Northern population
77% of the population live on North Island

Hatepe eruption

The eruption of Lake Taupo around AD 180, is known as the Hatepe eruption, the world's most violent in the past 5,000 years. It turned the skies over Europe red

Lake Taupo
2nd largest freshwater lake in Oceania, formed by the Oruanui volcanic eruption, the last supereruption on Earth, around 26,500 years ago. Islands 1,000 km away were covered with an 18-cm layer of ash

NEW ZEALAND
Named by Dutch explorers after the province of Zeeland in the Netherlands

Female suffrage
The 1st country to give women the right to vote, in 1893

Languages
New Zealand has three official languages: English, Māori and New Zealand Sign Language

TASMAN SEA

Around 1,500 km
to Australia

Most distant capitals
Wellington and Canberra are the two most distant capitals of neighbouring countries in the world, 2,326 km apart

New Zealand's ethnic groups

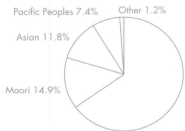

Pacific Peoples 7.4%
Other 1.2%
Asian 11.8%
Maori 14.9%
European New Zealander 64.7%

Kiwi
It's the only bird in the world to have a sense of smell, and of all birds, it lays the largest eggs in relation to its body size

The kiwi is the country's famous national symbol, prominent in the logo of the Royal New Zealand Air Force and the dollar coin

WELLINGTON

Southernmost capital
At latitude 41°18' S, it is the southernmost state capital on Earth

Cook Strait

Georgina Beyer
Became the world's 1st openly transsexual mayor in 1995, and Member of Parliament of New Zealand in 1999. She is one of the very few former sex-workers who has ever held public office

Aoraki / Mount Cook (3,724 m)
The tallest mountain in New Zealand. After 2014, it was listed as 40 m shorter due to rockslides and erosion

SOUTH ISLAND
(TE WAIPOUNAMU)

Southern Alps

Giant weta
Heaviest insect on the planet. The largest documented specimen weighed 71g

Animal kingdom
People constitute only 5% of the country's inhabitants, while animals represent the remaining 95%

PACIFIC OCEAN

Due to its remoteness, New Zealand was one of the world's last settled lands

Kiwifruit
Originated in China and was originally called Chinese gooseberry. New Zealand became one of its largest exporters. It first became popular among the US soldiers stationed in New Zealand during WWII. Farmers began calling it kiwi in the 1960s to increase its market appeal

Baldwin Street
The steepest residential street in the world with a 19° slope at its maximum

DUNEDIN

AUCKLAND ISLANDS

CHATHAM ISLANDS

Around 2,500 km
to Antarctica

STEWART ISLAND

| 0 | 50 | 100 | 200 k |

| 0 | 30 | 60 | 120 mi |

Arrival of humans
For over 50,000 years people have inhabited the highlands of New Guinea. They arrived here in one of the first migration waves of humankind, from Africa through Southeast Asia

Cassowary
One of New Guinea's predators and the 2nd heaviest bird in the world. This flightless bird weighs around 50 kg and is equipped with 12 claws. Fatal attacks on humans have been known

PACIFIC OCEAN

Biodiversity
New Guinea is home to about 10% of the world's vertebrates and 7% of all vascular plants grow here

Mystery
Papua New Guinea is one of the least geographically and culturally explored countries in the world. Many mountains and forests remain unnamed and unmapped

NORTHERN MARIANA ISLANDS (U.S.A.)

First humans in the Pacific
Tinian was one of the first inhabited islands in the Pacific. When people arrived here over 3,500 years ago, they had completed the longest seaborne journey humans had undertaken at that time

CAPITOL HILL

Latte Stones
These mushroom-shaped stones, weighing several tonnes, were raised by the ancient Chamorro people. It is believed that they were used to support the houses of upper class residents in ancient Guam

GUAM (U.S.A.)   HAGÅTÑA

0   75   150   225 km
0   50   100   150 miles

MANUS ISLAND
Home to the emerald green snail, whose shells are harvested to be sold as jewellery

Remoteness
Around 82% of the population in Papua New Guinea still live in isolated rural communities

NEW IRELAND

BISMARCK SEA

Kuk Swamp
Archaeological site with evidence explaining how plant exploitation transformed into agriculture. Cultivation in Kuk started at a similar time to the first civilization in Mesopotamia, and developed without interruption for the next 7,000 years

NEW BRITAIN

Size
New Guinea is the 2nd largest island on the planet, not counting Australia as an island

PAPUA NEW GUINEA

BOUGAINVILLE ISLAND

Division
From 1884 to 1919 the island was divided between the Dutch, Germans and British

Araucaria
One of the tallest evergreen, tropical trees. It can reach 10 m in circumference and grow to heights similar to the Statue of Liberty

Matschie's tree-kangaroo
It lives in the treetops of Huon Peninsula, spending most of its life eating leaves. Similarly to other kangaroos, it carries its joeys in a pouch

SOLOMON SEA

Gulf of Papua

Queen Alexandra's birdwing
Found only in Oro Province, it's an endangered species and the largest butterfly on the planet with a wingspan of 25 cm

Languages
There are over 850 languages spoken in Papua New Guinea, the highest number of any country

Torres Strait

PORT MORESBY

No megafauna
New Guinea's largest predators are the New Guinea singing dogs, inhabiting remote mountain regions and known for their distinctive, melodious howl

Isolated tribes
Until the end of the 20th century, more than 40 tribes living in a dense mountainous jungle in New Guinea had never been in contact with the outside world

Kokoda track
Historical and challenging 96-km trekking trail crossing the rainforest inhabited by the Koiari people. It was first used by European miners who sought access to the Yodda Kokoda goldfields. In WWII it became the site of a series of battles between Japanese and Australian forces

Independence
Papua New Guinea became independent from Australia in 1975

Hooded pitohui
It's one of the very few poisonous bird species. Just touching its feathers causes numbness

Battle of the Coral Sea
One of the major WWII battles of the Pacific, fought for 5 days in 1942 between Japan and the US and Australia. This was the first battle in which aircraft carriers engaged each other and played a major role in the fighting

CORAL SEA

75   150   300 km
50   100   200 miles

132

"Castle Bravo"
The largest nuclear test ever conducted by the US. The bomb, dropped on Marshall Islands' Bikini Atoll in 1954, was 1,000 times more powerful than the Hiroshima bomb

Size of the Castle Bravo nuclear "mushroom" cloud, around 150 km in diameter

DELAP-ULIGA-DJARRIT

MARSHALL ISLANDS

Amelia Earhart
Howland Island was a stopover during her round-the-world flight, which she would never complete

POHNPEI
PALIKIR
Nan Madol
Regarded as the "Venice of the Pacific", it's an ancient sunken city, consisting of a series of artificial islands off Pohnpei

FEDERATED STATES OF MICRONESIA

Nan Madol civilization
It existed for around 1,600 years, until the early 17th century, but its history still remains a mystery

Nuclear tests
The US conducted 105 nuclear tests here between 1946–62. Over 20,000 people were affected and the US paid over $750 million in reparations

Abanuea & Tebua Tarawa
In 1999, these small islands disappeared below the sea

BAIRIKI

GILBERT ISLANDS

HOWLAND ISLAND (U.S.A.)

Obese nation
Nearly 95% of Nauru's 10,000 population is overweight with 40% prevalence of type 2 diabetes

Challenges of prosperity
In recent decades, Nauru has become rich from phosphate mining. Many have given up living their traditional lifestyle, and changed their diet towards "Western", imported foods

YAREN

NAURU
World's 3rd smallest country after Vatican City and Monaco

Gilbert Islands
Home to over 90% of Kiribati nationals, with around 40% living on Tarawa atoll, which constitutes roughly 4% of the country's total land area

Caves of Nanumanga
It is believed that this underwater cave system, hidden 40 m down a coral cliff, had been used by people thousands of years ago when the sea level was much lower

Marovo Lagoon
Stretching to a length of 100 km, it's the world's largest saltwater lagoon

Natural blond
Around 10% of the native population has blond hair. This is not the same gene prevalent in northern Europe, but appeared in Oceania independently and was not passed down from European settlers

TUVALU

Flattened island
Funafuti island is less than 5 m above sea level because the US army flattened it during WWII to use as an airfield

VAIAKU

SOLOMON ISLANDS

HONIARA

Lake Tegano
Once a lagoon, it's the largest lake in the South Pacific

Transsexuals are widely accepted in Samoa and are considered a 3rd gender

WALLIS AND FUTUNA (France)

Kava-kava
Traditional plant consumed in a drink. It induces a state of deep relaxation and calmness. For centuries it's been used for medical and religious purposes. Even today, political figures visiting Vanuatu, are welcomed with kava

Discovery
Discovered in 1606 by a Portuguese explorer who thought he had found the hypothetical continent Terra Australis. Europeans did not return here until 1768

Fiji consists of over 330 islands, only 1/3 of which are inhabited. They are covered with tropical forests and mountains up to 1,300 m high

FIJI

VANUATU
PORT VILA

"The last eaten human"
Reverend Thomas Baker was killed and eaten for touching an aboriginal leader's head in 1867. The remains of his shoe are displayed in the Fiji Museum

SUVA

New Caledonian crow
It can select, process and combine twigs and leaves so that they serve a given function. These birds are the only non-primate species capable of making and using complex tools

NEW CALEDONIA (France)

Isle of Pines
Captain James Cook was amazed when he saw the island filled with 50 m high pine trees instead of palms

"You are welcome"
Fiji's government reacted delightfully to the threat of rising sea levels and possible forced migration of Pacific nations: "If you need a place to live, you can come and live here"

0 150 300 600 km
0 100 200 400 miles

Pacific islands
There are 4 different types of islands in the Pacific: continental islands, coral reefs, volcanic islands and raised coral platforms

PALMYRA ATOLL (U.S.A.)

Guano
The Howland, Baker and Jarvis islands were rich in guano – accumulated excrement of seabirds, used as a fertilizer. It was mined by American companies between 1857 and 1879

BAKER ISLAND (U.S.A.)

KIRITIMATI
It's the largest atoll in the world, making up around half of the entire country's land area

JARVIS ISLAND (U.S.A.)

KIRIBATI
Consists of 33 atolls and reef islands, 12 of which are uninhabited

Disappearing world
The islands of Kiribati have been inhabited by the people of Micronesia since 3000 BC, however it is most likely that they will perish in our lifetime

PACIFIC OCEAN

Geographically stretched
Although only around 112,000 people live in Kiribati, its area is spread across 4 different time zones over 3,000 km

"Low islands"
The altitude of the Kiribati islands averages about 1–2 m, with the highest point just 81 m above sea level

TOKELAU (N.Z.)

Society of Tokelau
Despite their national identity and urging from the UN, the people of Tokelau have voted twice to remain a dependency of New Zealand

Climate-induced migration
In 2020, the people of Kiribati will start a nationwide relocation to Fiji where they have purchased habitable land, less exposed to submerging

Tourism accounts for 25% of Samoa's GDP

During the 1918 flu pandemic, American Samoa was one of only a few flu-free places in the world

The 1st solar nation
With a population of 1,250, Tokelau is the world's first 100% solar-powered nation

Surfing
Originally developed by the Polynesians, it was first noted during the first voyage of Captain James Cook in 1769

SAMOA APIA

Nonpartisanism
In Tokelau's elections, people vote for individuals and not for political parties

Tahitian pearl
Black pearls. Constitute around 55% of French Polynesia's exports

FRENCH POLYNESIA (France)

AMERICAN SAMOA (U.S.A.)
Southernmost territory of the USA

TAHITI

Teahupo'o
One of the world's most famous and epic surfing spots with waves up to 7 m

TONGA
Constitutional monarchy. The current ruler, King Tupou VI, was crowned in 2015

COOK ISLANDS (N.Z.)
The archipelago of 15 islands was named by the Russians in the early 1820s in honour of Captain James Cook

Tattoo
The word originates from the Polynesian language and the art plays a vital part in the culture of indigenous Polynesians

Tautira Bay
Since its discovery in 1772, its landscape has served as inspiration for artists, writers and explorers such as Paul Gaugin, Robert Louis Stevenson and Henry Adams

"Friendly Islands"
Named for the congenial reception accorded to Captain James Cook on his first visit

NUKU'ALOFA

Horizon Deep
Deepest point of the Tonga Trench and the 2nd deepest place in the oceans (-10,800 m)

Tiare flower
National symbol of Tahiti, worn by both men and women. You wear it on your right ear to show that you're single, or on your left ear if you're in a relationship

Megatsunami

It is possible that an eruption of the volcano on the island of La Palma in the Canary Islands could cause part of the volcano to slide into the ocean and trigger a 50 m megatsunami that would hit North America

ANTARCTICA & OCEANS

First base ★
First meteorological station established by the Scottish National Antarctic Expedition in 1903. 36 m², it had two windows and could fit 6 men

Endurance expedition

Drake Passage

Expedition launched boats

Chile/Argentina/U.K.

WEDDELL SEA

Flight to Antarctica

"San Telmo" sinking ★
644 passengers of this Spanish ship are probably the first known people to die in Antarctica (1819)

Endurance expedition
Failed attempt to make the first land crossing of the Antarctic continent

"Endurance" sank
"Endurance" crushed by ice

Ship "Endurance" trapped in ice

High elevation
Antarctica has the highest average elevation (2,300 m) of all the continents

Flight to Antarctica
Taking a plane ($5,000–10,000) from Chile is the easiest and fastest way to get to Antarctica

Ice shelves
Floating ice platforms that form where a glacier meets the ocean. The majority of· Antarctic ice cap loss is caused by the melting of the ice shelves, which then break into mega icebergs

ANTAI

Ronne Ice Shelf

Mount Tyree (4,852 m) ★
Only 10 people have ever reached the summit

★ Vinson (4,897 m)
Highest mountain in Antarctica. First ascent in 1966. 1,200 climbers have reached the summit successfully. No deaths have ever been reported. Total expedition cost is around $30,000

Nesting coast
During the Polar spring, the Antarctic coast provides a nesting place for more than 100 million birds

Pine Island Glacier
The largest and the fastest melting glacier on the continent, responsible for 1/4 of its ice loss. A rift formed at the base of the ice shelf which worked its way up for 2 years until it caused a 580 km² iceberg to break off in 2015

★ Bentley Trench (-2,555 m)
It's the deepest point on the planet not covered by the ocean, although covered by ice

West Antarctica
·The West ice sheet is much less stable because it rests on land lying mostly below sea level

Ross Ice Shelf
Size of France

Amundsen 14.12.1911

AMUNDSEN SEA

Global warming
CO_2 A 1°C temperature rise in the oceans will lead to the ice shelves melting from below

"The Last Ocean"
Ross Sea is one of the most pristine marine ecosystems which has remained free from pollution and overfishing. The most productive stretch of water is in the Southern Ocean and is inhabited by large predatory fish, whales, seals and penguins

ROSS SEA

Unclaimed

Mount Erebus (3,794 m)
Earth's southernmost active volcano and second highest volcano in Antarctica

SOUTHERN OCEAN

Antarctic Treaty System
Prohibits military activities, mineral mining, nuclear explosions and nuclear waste disposal, and holds all territorial claims in abeyance

McMurdo Station
US research centre, the largest in the Antarctic. Can support up to 1,258 residents. Anyone wanting to leave the station has to attend a survival course

| 0 | 250 | 500 | 1,000 km |

| 0 | 150 | 300 | 600 miles |

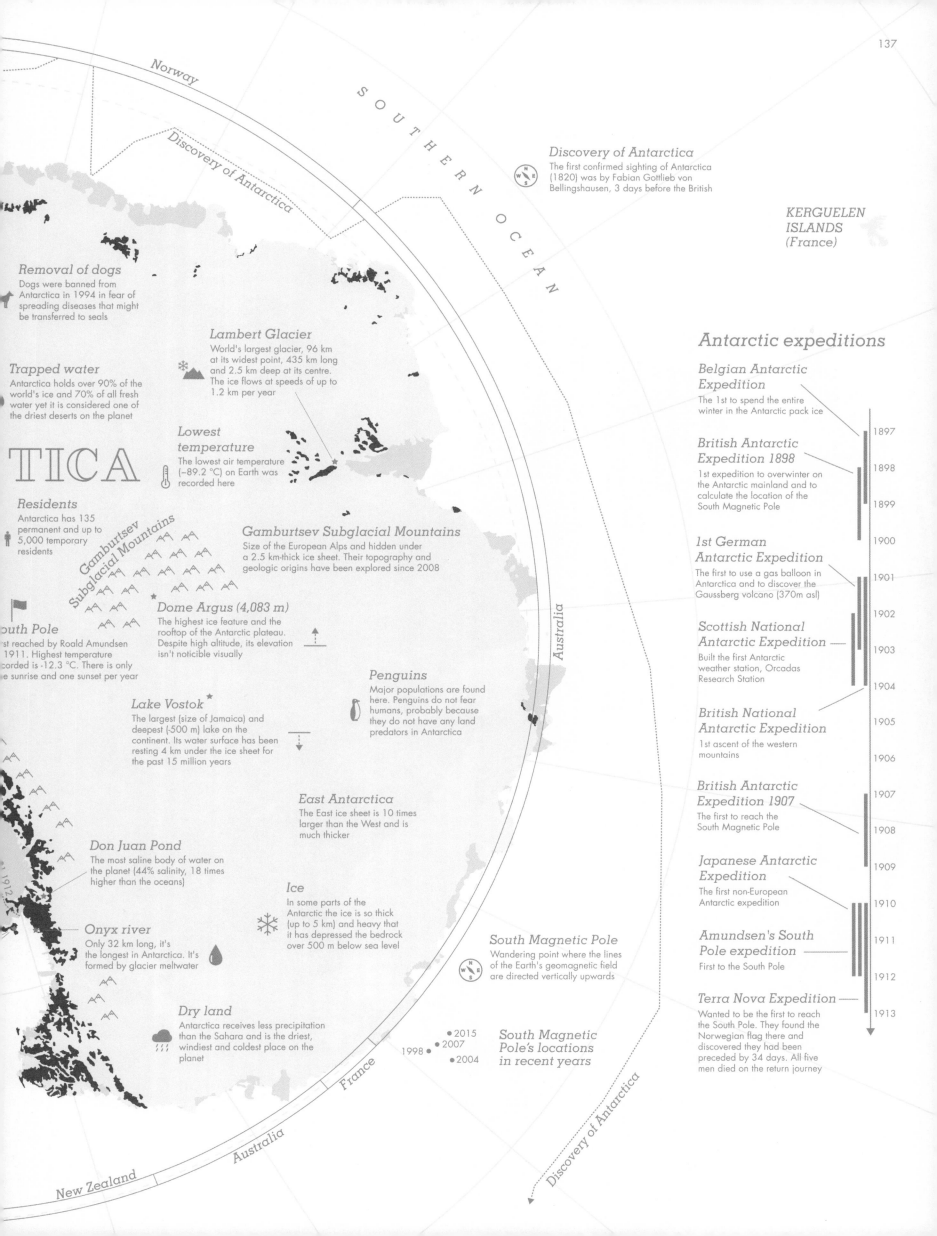

S O U T H E R N   O C E A N

Norway

Discovery of Antarctica

Discovery of Antarctica
The first confirmed sighting of Antarctica (1820) was by Fabian Gottlieb von Bellingshausen, 3 days before the British

KERGUELEN ISLANDS
(France)

Removal of dogs
Dogs were banned from Antarctica in 1994 in fear of spreading diseases that might be transferred to seals

Lambert Glacier
World's largest glacier, 96 km at its widest point, 435 km long and 2.5 km deep at its centre. The ice flows at speeds of up to 1.2 km per year

Trapped water
Antarctica holds over 90% of the world's ice and 70% of all fresh water yet it is considered one of the driest deserts on the planet

Lowest temperature
The lowest air temperature (–89.2 °C) on Earth was recorded here

Antarctic expeditions

Belgian Antarctic Expedition
The 1st to spend the entire winter in the Antarctic pack ice

TICA

Residents
Antarctica has 135 permanent and up to 5,000 temporary residents

Gamburtsev Subglacial Mountains

Gamburtsev Subglacial Mountains
Size of the European Alps and hidden under a 2.5 km-thick ice sheet. Their topography and geologic origins have been explored since 2008

British Antarctic Expedition 1898
1st expedition to overwinter on the Antarctic mainland and to calculate the location of the South Magnetic Pole

1897

1898

1899

1st German Antarctic Expedition
The first to use a gas balloon in Antarctica and to discover the Gaussberg volcano (370m asl)

1900

1901

...uth Pole
...st reached by Roald Amundsen ...1911. Highest temperature ...corded is -12.3 °C. There is only ...e sunrise and one sunset per year

Dome Argus (4,083 m)
The highest ice feature and the rooftop of the Antarctic plateau. Despite high altitude, its elevation isn't noticible visually

Scottish National Antarctic Expedition
Built the first Antarctic weather station, Orcadas Research Station

1902

1903

1904

Lake Vostok
The largest (size of Jamaica) and deepest (-500 m) lake on the continent. Its water surface has been resting 4 km under the ice sheet for the past 15 million years

Penguins
Major populations are found here. Penguins do not fear humans, probably because they do not have any land predators in Antarctica

British National Antarctic Expedition
1st ascent of the western mountains

1905

1906

East Antarctica
The East ice sheet is 10 times larger than the West and is much thicker

British Antarctic Expedition 1907
The first to reach the South Magnetic Pole

1907

1908

Don Juan Pond
The most saline body of water on the planet (44% salinity, 18 times higher than the oceans)

Ice
In some parts of the Antarctic the ice is so thick (up to 5 km) and heavy that it has depressed the bedrock over 500 m below sea level

Japanese Antarctic Expedition
The first non-European Antarctic expedition

1909

1910

Onyx river
Only 32 km long, it's the longest in Antarctica. It's formed by glacier meltwater

South Magnetic Pole
Wandering point where the lines of the Earth's geomagnetic field are directed vertically upwards

Amundsen's South Pole expedition
First to the South Pole

1911

1912

Dry land
Antarctica receives less precipitation than the Sahara and is the driest, windiest and coldest place on the planet

• 2015
• 2007
1998 •   • 2004

South Magnetic Pole's locations in recent years

Terra Nova Expedition
Wanted to be the first to reach the South Pole. They found the Norwegian flag there and discovered they had been preceded by 34 days. All five men died on the return journey

1913

Australia

France

New Zealand

Australia

Discovery of Antarctica

Drainage
Only around 13% of
the world's land
drains into the Pacific

Japanese spider crab
It has the greatest leg span of any
arthropod (up to 5.5 m)

Saline water
The average salinity of
the oceans is 3.5%. The
higher it is, the lower its
freezing point

Xu Fu
He was a Chinese court
sorcerer of the 3rd century BC
who explored the Pacific
searching for the elixir of life

Surrender of Japan
Signed on 2 September
1945 in Tokyo Bay aboard
the USS *Missouri*

Volume
The Pacific represents 50%
of the world's oceanic water
volume, around 714 million km³

Challenger Deep
Deepest point on Earth, between
10,898 and 10,916 m deep. Only
3 explorers have visited it: Jacques
Piccard and Don Walsh in 1960
and James Cameron, the film
director, in 2012

The pressure on its floor is 1,000 times
that on the surface. Sunlight doesn't
reach there, making photosynthesis
impossible. Water temperature
is just above freezing

Size
The Pacific covers around 51% of the world's
oceans area and more than 30% of the Earth's
total surface area. All of the continents
combined could fit within its boundaries

Depth
Average depth of the
Pacific is 4,200 m, more
than any other ocean

Pacific War
Named after the theatre of WWII,
fought in East Asia and the Pacific
Ocean from 1941 to 1945

Mariana Trench

Greatest width

Giant Pacific octopus
The largest octopus on the planet
can grow up to 3.3 m long

Sound in water
Sound travels 4.3 times
faster in water than in air

PACIFI

Maris Pacifici
First detailed map of the Pacific,
created by Abraham Ortelius in
1589, 19 years after he created
Theatrum Orbis Terrarum, the first
true modern atlas

Pacific islands
There are over 25,000 islands in the Pacific.
If one wished to visit all the Pacific islands
and spend at least one day on each of them,
it would require nearly 70 years of travel

NEW ZEALAND
New Zealand was discovered
in 1642 by the Dutchman
Abel Tasman, after whom the
sea and the island were named

**Centre of the
water hemisphere**
Centre of the Earth's
hemisphere containing the
largest possible area of water

Asia-Pacific Economic Cooperation
APEC is a forum of 21 countries bordering the
Pacific, promoting free trade throughout the region.
Traditionally, all attending heads of government
dress in the national costume of the host country

0	625	1,250	2,500 km
0	375	750	1,500 miles

Black smoker chimneys
If life originated in the oceans, it is most likely to have happened in such volcanic vents, emitting water filled with minerals, heated by the Earth's crust

Countries
42 sovereign nations border the Pacific

Seven Seas
Ancient term for all of the world's oceans: Arctic, North Atlantic, South Atlantic, Indian, North Pacific, South Pacific and Southern

Origin of life
Despite the temperature near the black smokers exceeding 400 °C, the simplest form of life lives there – a bacteria which turns geothermal water into simple sugars at zero-sunlight conditions. Shrimps, tube worms and other creatures feed on the bacteria, initiating the entire oceans' food cycle

Port of Long Beach
The largest US port by total trade on the Pacific is only its 5th largest port

Pearl Harbor
Base for the US Pacific Fleet consisting of 250,000 personnel, 2,000 aircraft and 200 ships

Longest sea survival
In 2012, two men went fishing, got caught in a storm and drifted 10,700 km for 438 days. They lived off fish and birds and drank turtles' blood. Only one of them survived

First sight
The eastern Pacific was first reached by Europeans in 1513 by a Spanish explorer, Vasco Núñez de Balboa, who crossed the Isthmus of Panama by land

Longest sea survival

Greatest width
The Pacific Ocean reaches its greatest east-west width (19,800 km) at around 5°N latitude, between Colombia and Indonesia

Division
The equator divides the ocean into the North Pacific and South Pacific

GALAPAGOS ISLANDS (Ecuador)
Located very close to the centre of the Western Hemisphere

CEAN

Kon-Tiki expedition

Kon-Tiki expedition
Thor Heyerdahl, a Norwegian adventurer, hypothesised that the indigenous cultures from South America could have spread throughout the Pacific in pre-Columbian times. In 1947, he travelled 6,900 km on a primitive raft to prove the theory. The 1950 Academy Award-winning documentary "Kon-Tiki" tells his story

War of the Pacific
The Chilean victory over Bolivia and Peru in this 1879–83 war significantly expanded its territories

Plastic
Plastic pollution of the Pacific increased a hundredfold from 1972 to 2012

EASTER ISLAND (Chile)
The most remote airport in the world – it is located 2,603 km from Gambier Islands' airport

Pacific Alliance
Latin American trade bloc bringing together Chile, Colombia, Mexico and Peru

Mystery
Today's science has in fact better knowledge of the surface of Mars than about the oceans' floor

Spacecraft cemetery
Located near Point Nemo. Between 1971 and 2016, over 263 spacecraft were disposed of in this area

★ Point Nemo
Oceanic point of inaccessibility – the most isolated place, 2,688 km from the nearest land

Elephant seal
There are only two species of elephant seal left – northern and southern. Both were hunted to near-extinction until the late 19th century

Naming
When Ferdinand Magellan first reached the Pacific in 1521 during his attempt to circumnavigate the Earth, it seemed to him much calmer than the Atlantic, and thus named it "mar pacifico"

"The Iceberg"
The iceberg that hit the Titanic was probably formed 3 years earlier. The part visible above water may have been the size of the Roman Colosseum. The remaining 90% of ice-mass was hidden below the surface of the water

1st transatlantic cable
Laid in 1858. The first telegram was a message of congratulations from Queen Victoria to the US President James Buchanan. Without the cable a letter would have taken 10 days to travel by ship

Bermuda Triangle
Site of numerous aviation and shipping incidents as a result of unexplained and supposedly mysterious causes

1st transatlantic flight
In 1919, a British aeroplane flew 3,040 km in 15 h 57 min non-stop across the Atlantic, the first to do so

"Green Atlantic"
The term describes the 4.6 million Irish who migrated to the US between 1820 and 1930

1st transatlantic flight

1st transatlantic cable

Titanic

Sinking of the Titanic ★
The ship sank 4 days into the voyage, resulting in over 1,500 deaths. Its construction had advanced safety measures, but it lacked lifeboats for half of the passengers

Cuvier's beaked whale
Can reach depths of 3,000 m, deeper than any other mammal

SARGASSO SEA
The only sea that is bounded by currents and not by land. It is largely covered with seaweed that has been floating there for millions of years

Icebergs
Have been spotted as far south as 32° N, a similar latitude to Madeira

Trade
The volume of Mediterranean trade was surpassed by that of transatlantic trade by the end of the 17th century

Sargassum fish
Fish species endemic to the Sargasso Sea, originating from what is today the Carpathian region, from where it migrated 17 million years ago

Megatsunami
It is possible that an eruption of the volcano on the island of La Palma in the Canary Islands could cause part of the volcano to slide into the ocean and trigger a 50 m megatsunami that would hit North America

Raft crossing

Raft crossing
In 1984, five Argentines sailed a hand-made raft for 52 days across the Atlantic, to prove that Africans might have crossed to America much earlier than Columbus did

"Black Atlantic"
The term refers to the Atlantic slave trade and its influence on America and Africa

Size
2nd largest ocean, covering around 20% of the Earth's total surface and around 29% of its water surface area

Salinity
The saltiest of all oceans

ATLANTIC OCEAN

0°, 0° ★
The "centre" of the Earth's surface according to the standard geographic model

Planet's drainage
Even although it's not the largest ocean, around 49% of the Earth's land drains to the Atlantic

Aethiopian Sea
Classical name used to describe the southern part of the Atlantic, until the 19th century

ASCENSION (U.K.)
The Global Positioning System, uses 31 satellites and 6 monitor stations — one of them is located here

ST HELENA (U.K.)
Napoleon was exiled here in 1815 by the British and died 6 years later

Mid-Atlantic Ridge
Running through the middle of the ocean, it is a part of the world's longest mountain range

The two most remote permanently inhabited places lie 2,434 km apart

Jellyfish
World's oldest multi-organ animal. It originated around 550 million years ago. It has no brain and 90% of its body is composed of water

★ TRISTAN DA CUNHA (U.K.)

Blue whale
The largest animal on Earth. At around 170 tonnes and up to 30 m in length, it consumes krill almost exclusively. Hunted almost to extinction

BOUVET ISLAND (Norway)
The most remote island in the world, over 1,600 km to the nearest land (Antarctica)

0	625	1,250	2,500 km
0	375	750	1,500 miles

Largest rivers
The largest rivers by discharge that flow into the Indian Ocean are the Ganges-Brahmaputra, Zambezi and Irrawaddy which are only 3rd, 22nd and 41st largest in the world respectively

Strategic choke points
Strait of Hormuz, Strait of Malacca, Suez Canal, Bab al Mandab and Cape of Good Hope are amongst the world's most important choke points

★ Suez Canal

Strait of Hormuz

Ganges-Brahmaputra

Irrawaddy

Petroleum
Ocean's most valuable mineral resource. Around 40% of global oil production comes from this ocean, mostly from The Gulf

Indian Ocean trade
Busiest trade routes connecting East Asia and The Gulf with Europe cross the Indian Ocean. The Suez Canal reduced the distance by sea by 40%

Centre of population
The place closest to all humans on the planet is located somewhere on the Indian subcontinent but its exact location is constantly shifting

Bab al Mandab

Early explorations
Egyptians probably began exploring the Ocean around 2,300 BC

Indian Ocean trade

Strait of Malacca

High temperature
Compared to other oceans, it has a limited marine life due to high water temperature

★ **Carlsberg Ridge**
Named after the Carlsberg Group brewery which sponsored the 1928–30 expedition that discovered it

2004 Tsunami ★
An undersea earthquake triggered a series of devastating tsunamis along the coasts of 14 countries killing up to 280,000 people (mostly Indonesians) with waves of up to 30 m high

British Indian Ocean Territory
Location of a joint military base of the UK and the US, one of a few Western military bases in the region

Monsoon
The Indian Ocean's currents are mainly controlled by the monsoon reversal every half year

Malaysia Airlines Flight 370
In 2014, a Boeing 777 went missing in unexplained circumstances killing 239 people. The largest and most expensive search in the history of aviation turned up no sign of wreckage but some debris, found on the coast of Africa and Indian Ocean islands nearby, have been confirmed as from the aircraft

Zambezi

INDIAN OCEAN

Leatherback sea turtle
The largest of all turtles, it can weigh up to 700 kg. It lacks a bony shell, typical of other turtles

Uninhabited islands
Unlike the Pacific, almost all of the Indian Ocean's islands were uninhabited until colonial times

Size & depth
3rd largest ocean, covering around 22% of the water on the Earth's surface and 2nd deepest with an average depth of around 4,000 m

Portuguese hegemony
From the beginning of the 16th century until the mid 17th century, the ocean was dominated by the Portuguese

Garbage patch
Discovered in 2010, this patch is 5 million km² and circulates the ocean in a 6-year cycle. Over 6 billion kg of plastic is dumped into the oceans yearly. If this continues, in 30 years there could be more plastic than fish in the sea

Manganese nodules
These rock "balls" that lie on the sea bottom (mostly in the Indian Ocean) are rich in valuable metals

"Marine snow"
Continuous fall of organic material from the highest layers of the ocean, consisting of the remains of dead organisms. It gives life to other organisms living in deep darkness

Cape of Good Hope

CROZET ISLANDS (France)
In 1887, a French ship was wrecked here. Its survivors tied an SOS note to the leg of an albatross. The note was found 7 months later in Australia but too late to save anyone

KERGUELEN ISLANDS (France)
One of the most isolated inhabited places on Earth

Formation
The Indian Ocean was formed after the breakup of the Gondwana supercontinent, 180 million years ago, when the Indian subcontinent moved northeast

Fishing
Fishing boats from Asian countries of the Pacific exploit the Indian Ocean for shrimp and tuna

Southern Ocean
Waters south of 60°S are considered part of the Southern Ocean

| 0 | 625 | 1,250 | 2,500 km |
| 0 | 375 | 750 | 1,500 miles |

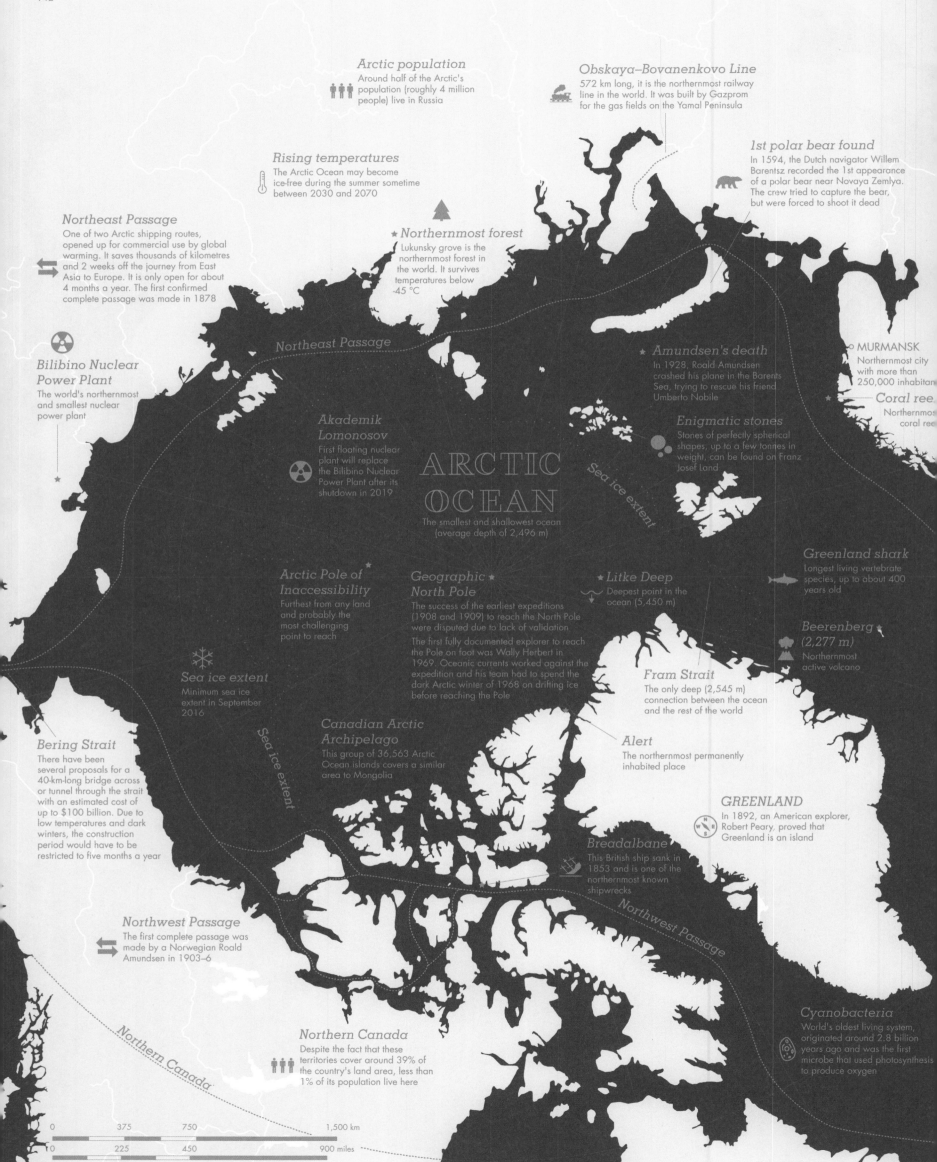

Arctic population
Around half of the Arctic's population (roughly 4 million people) live in Russia

Obskaya–Bovanenkovo Line
572 km long, it is the northernmost railway line in the world. It was built by Gazprom for the gas fields on the Yamal Peninsula

1st polar bear found
In 1594, the Dutch navigator Willem Barentsz recorded the 1st appearance of a polar bear near Novaya Zemlya. The crew tried to capture the bear, but were forced to shoot it dead

Rising temperatures
The Arctic Ocean may become ice-free during the summer sometime between 2030 and 2070

★ **Northernmost forest**
Lukunsky grove is the northernmost forest in the world. It survives temperatures below -45 °C

Northeast Passage
One of two Arctic shipping routes, opened up for commercial use by global warming. It saves thousands of kilometres and 2 weeks off the journey from East Asia to Europe. It is only open for about 4 months a year. The first confirmed complete passage was made in 1878

Northeast Passage

★ **Amundsen's death**
In 1928, Roald Amundsen crashed his plane in the Barents Sea, trying to rescue his friend Umberto Nobile

MURMANSK
Northernmost city with more than 250,000 inhabitants

Coral reef
Northernmost coral reef

Bilibino Nuclear Power Plant
The world's northernmost and smallest nuclear power plant

Akademik Lomonosov
First floating nuclear plant will replace the Bilibino Nuclear Power Plant after its shutdown in 2019

Enigmatic stones
Stones of perfectly spherical shapes, up to a few tonnes in weight, can be found on Franz Josef Land

Sea ice extent

ARCTIC OCEAN
The smallest and shallowest ocean (average depth of 2,496 m)

Greenland shark
Longest living vertebrate species, up to about 400 years old

★ **Litke Deep**
Deepest point in the ocean (5,450 m)

Arctic Pole of Inaccessibility
Furthest from any land and probably the most challenging point to reach

Geographic North Pole
The success of the earliest expeditions (1908 and 1909) to reach the North Pole were disputed due to lack of validation

The first fully documented explorer to reach the Pole on foot was Wally Herbert in 1969. Oceanic currents worked against the expedition and his team had to spend the dark Arctic winter of 1968 on drifting ice before reaching the Pole

Beerenberg (2,277 m)
Northernmost active volcano

Sea ice extent
Minimum sea ice extent in September 2016

Sea ice extent

Fram Strait
The only deep (2,545 m) connection between the ocean and the rest of the world

Alert
The northernmost permanently inhabited place

Bering Strait
There have been several proposals for a 40-km-long bridge across or tunnel through the strait with an estimated cost of up to $100 billion. Due to low temperatures and dark winters, the construction period would have to be restricted to five months a year

Canadian Arctic Archipelago
This group of 36,563 Arctic Ocean islands covers a similar area to Mongolia

GREENLAND
In 1892, an American explorer, Robert Peary, proved that Greenland is an island

Breadalbane
This British ship sank in 1853 and is one of the northernmost known shipwrecks

Northwest Passage
The first complete passage was made by a Norwegian Roald Amundsen in 1903–6

Northwest Passage

Northern Canada

Northern Canada
Despite the fact that these territories cover around 39% of the country's land area, less than 1% of its population live here

Cyanobacteria
World's oldest living system, originated around 2.8 billion years ago and was the first microbe that used photosynthesis to produce oxygen

| 0 | 375 | 750 | 1,500 km |
| 0 | 225 | 450 | 900 miles |

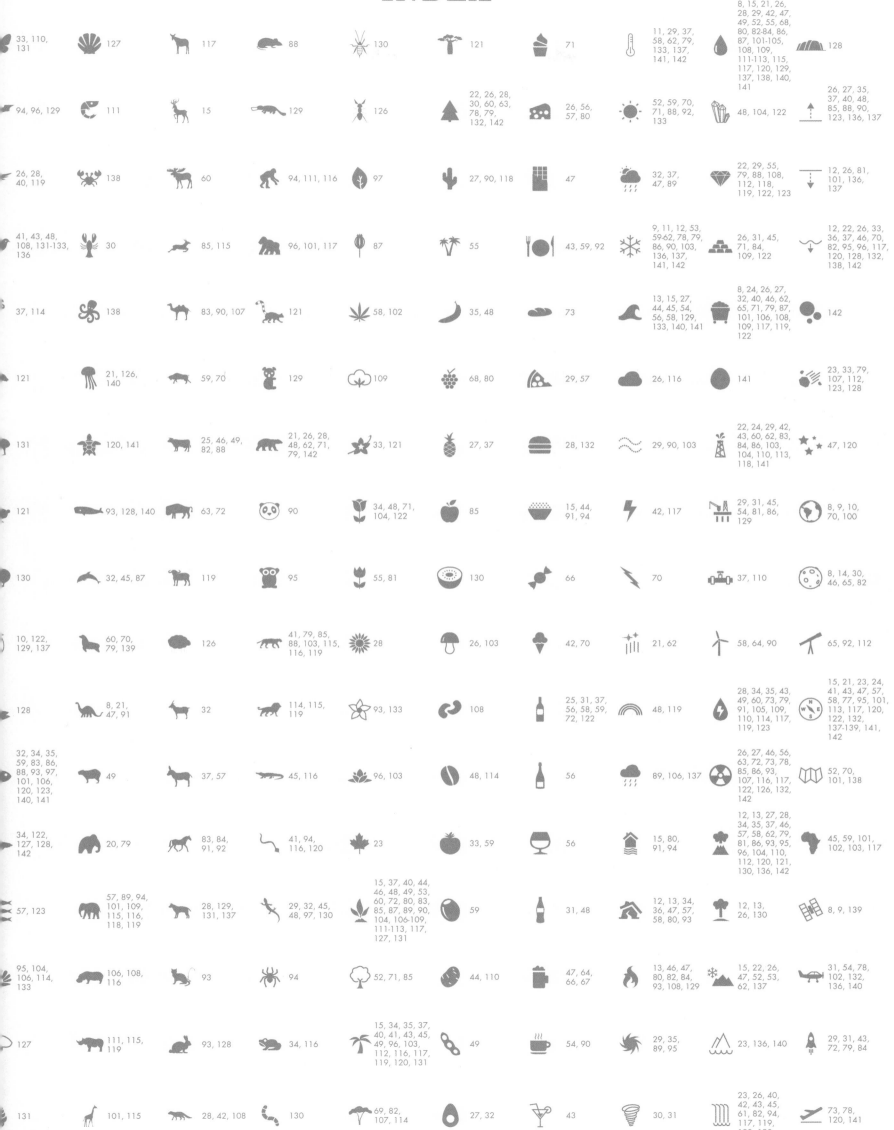

30, 31, 60 · 26, 63, 72 · 72 · 24, 61, 63, 92 · 47, 67 · 11, 26, 30, 32, 36, 45, 47, 57, 65, 72, 79, 89, 112 · 73 · 36

30, 32, 54, 55, 100, 112, 130, 139, 140 · 72, 73, 79, 85 · 58, 76, 77, 81, 83, 96, 102, 107, 108, 109, 111, 113 · 91 · 35, 107 · 56, 60, 82, 103 · 28, 52, 64, 78, 90 · 11, 42, 73, 93, 101, 114, 115 · 126

10, 11, 61, 70, 83 · 25, 29, 30, 33, 34, 41, 42, 45, 48, 66, 92, 117 · 11, 24, 36, 41, 42, 49, 54, 58, 64, 69, 71, 76, 81, 82, 89, 97, 105, 109, 114, 127 · 93 · 87 · 67 · 70 · 71 · 86

10, 47, 52, 57, 80, 96, 106, 136, 141, 142 · 56, 64, 69, 73, 78, 81, 83, 87, 89, 94, 102, 110, 111, 113, 114 · 61, 65, 71, 72, 82 · 91 · 136, 140 · 47, 105 · 15, 105, 113 · 36, 49, 66, 89 · 61, 77, 104, ...

20, 25, 54, 140 · 29, 30, 30, 31, 49 · 88, 121 · 한글 92 · 55, 66 · CO_2 13, 33, 90, 111, 136 · 71, 94, 104 · 89, 115 · 29, 1...

23-27, 30, 41, 43, 44, 46, 56, 62, 64, 71, 73, 81, 100, 103, 109, 112, 128, 130 · 54, 55, 56, 59, 61, 65, 67, 71 · 76, 87, 89, 91, 94 · € 53, 59, 60, 69 · 23, 28, 30, 31, 33, 44, 54, 58, 61, 64, 88, 107 · 9, 139, 142 · 105 · 68 · 23, 65...

46, 106 · 70 · محمد 83 · ① 25, 70, 107, 121 · 24, 30, 87, 101, 133 · 54, 57, 80 · 103, 106 · 68 · 72

25, 30, 69, 71, 90, 114 · 23, 25, 26, 29, 34, 35, 37, 44, 45, 49, 60, 61, 64, 68, 72, 80, 83, 85, 86, 88, 89, 91, 92, 94, 96, 97, 100, 101, 103, 104, 107, 108, 114, 118, 120, 129, 131, 133, 137 · 109 · 61 · 32 · 15, 40, 48, 52, 56, 57, 70, 73, 77, 81, 82, 83, 88, 103, 105, 132 · 24, 32, 64, 102 · 24, 36, 54, 60, 68, 70, 71, 80, 91, 92, 114, 126 · 92, 10...

56, 57, 67 · 28, 58, 63, 80, 82, 83, 88, 105, 117, 118, 119, 121, 127, 138 · 56, 65 · 11, 15, 20, 21, 23, 24, 31, 35, 37, 48, 49, 53, 55, 63–65, 76, 77, 82, 86, 87, 90, 91, 105–107, 109, 114, 115, 117, 127, 128, 138-142 · 34, 36, 123 · 94 · 88 · 68 · 56, 64, 65, 82...

24, 30, 33, 54, 68, 79, 87, 89, 94, 109, 116, 120, 128, 142 · 65 · 59 · 9, 33, 41, 45, 47, 49, 63, 77, 88, 91, 92, 96, 110, 118, 119 · 29, 45 · 55 · 20, 22-26, 29, 30, 36, 40, 43, 45, 47, 52, 53, 55, 64, 71, 76, 77, 79, 82-84, 87-91, 93, 95-97, 100, 102, 106, 110, 111, 113, 115, 116, 119, 121, 126-128, 130-132, 137, 140-142 · 55, 63 · 71

56, 66, 93 · 36, 52, 68, 73, 78, 81, 84, 85, 92, 109, 118 · 28, 31, 56, 57, 86, 88, 115, 117, 120 · 53, 55, 80, 121 · 11, 21, 35, 36, 53, 59, 73, 76, 87, 114, 121 · 104 · 86 · 63, 81 · 91

55, 61, 116 · 11, 15, 30, 33, 47, 55, 58, 63, 67, 71, 73, 82, 84, 102, 114, 118, 132, 136, 139 · 45 · $ 24, 27, 31, 35, 36, 40, 42, 43, 45, 52, 59, 63, 76, 78, 83-85, 87, 91, 96, 100, 101, 107, 110, 115, 116, 126, 132 · 122 · 64 · 91, 106, 107, 112 · 25, 66, 76, 90, 131, 136 · 139, ...

28, 40, 94 · 28, 29, 41, 58, 100, 106, 108, 109, 140 · 57 · 42, 46, 70, 72, 112, 113, 117 · 67 · 105 · 24, 29, 30, 42, 44, 66, 78, 83, 85, 86, 92, 106, 109, 111, 115, 117, 130 · 10, 22, 66, 67, 103 · 92

25, 35, 36, 48, 55, 56, 63, 65, 69, 71, 73, 78, 80, 82, 86, 87, 92, 94, 103, 108, 110, 112, 113, 118, 119 · 55, 59, 63, 84, 92, 100, 116, 117, 119, 123 · 11, 28, 31, 46, 49, 54, 57, 58, 61, 64, 83, 107, 114, 117 · % 67 · 20, 40, 45, 83, 105 · 10, 56, 64, 67, 71, 72 · 97 · 9, 36, 101, 102, 114

25, 47, 48, 56, 61, 65, 69, 73, 92, 93, 120, 127, 131, 138, 139, 141 · 23, 30, 31, 45, 53, 54, 55, 56, 58, 60, 61, 64, 67, 68, 69, 77, 80, 82, 94, 96, 101, 106, 123, 128, 141 · 26, 28, 29, 30, 53, 55, 58, 59, 64, 65, 66, 93, 109 · 29 · 83, 108 · 83, 92 · 66, 106 · 87, 103, 111, 127

12, 24, 34, 64, 81, 132, 139 · 105 · 129 · 95 · 21, 70, 80, 93, 119, 123 · 33, 34 · 62, 66, 72, 83, 95 · 126

26, 92 · 10, 23, 62 · 15, 44, 89, 131 · 112 · 54, 65, 66, 76, 82, 87, 89, 94, 108, 109, 123 · 59 · 23, 37, 40, 43, 44, 49, 58, 59, 61, 67, 71, 73, 87, 96, 101, 107, 111, 114, 116, 122, 130, 131 · 22, 63

67 · 67 · 60 · 61, 90, 112 · 67 · 23, 93 · 55, 60, 62 · 32

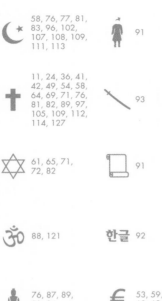